The Building Arts of South Texas

The Building Arts of South Texas:
Stories of Endangered Building Arts &
The Craftsmen Who Keep Them Alive

Published by
FRANKLIN SCRIBES
San Antonio, Texas

Email: info@franklinscribes.com

THE BUILDING ARTS GROUP
Emily Denman Thuss, Barbara Dean Hendricks,
Monte Adams, and Tommy Adams

A partnership created to document, through in-depth interviews and photography, some of the talented artisans working in San Antonio and South Texas.

Website: buildingartsgroup.com
Books available at Amazon.com

TRADECLOTH EDITION
ISBN-13: 9781941516065
Library of Congress Control Number: 2014959389

TRADEPAPER EDITION
ISBN-13: 9781941516072

Oral History/Building Crafts
Historical Preservation

Printed and bound in the United States of America
Book Design and Layout by Don Kirk

Thanks go out to the craftsmen in this book who sat for hours on end as I listened intently to each of their fascinating life stories.
—Barbara Dean Hendricks

06v10

To Ron and Nancy —
Enjoy the History!

The Building Arts of South Texas

Stories of Endangered Building Arts & The Craftsmen Who Keep Them Alive

By

Barbara Dean Hendricks

Dedication

This book is dedicated to Emily Denman Thuss, whose vision, acumen, persistence, and generosity made this book possible.

We would also like to acknowledge the people who inspired Emily to develop this book: the grand, old-world craftsmen and the many generations of direct descendants who have taken up the trade in order to preserve and pass along the skills needed to restore our historic architecture so that future generations might have a chance to enjoy and appreciate it.

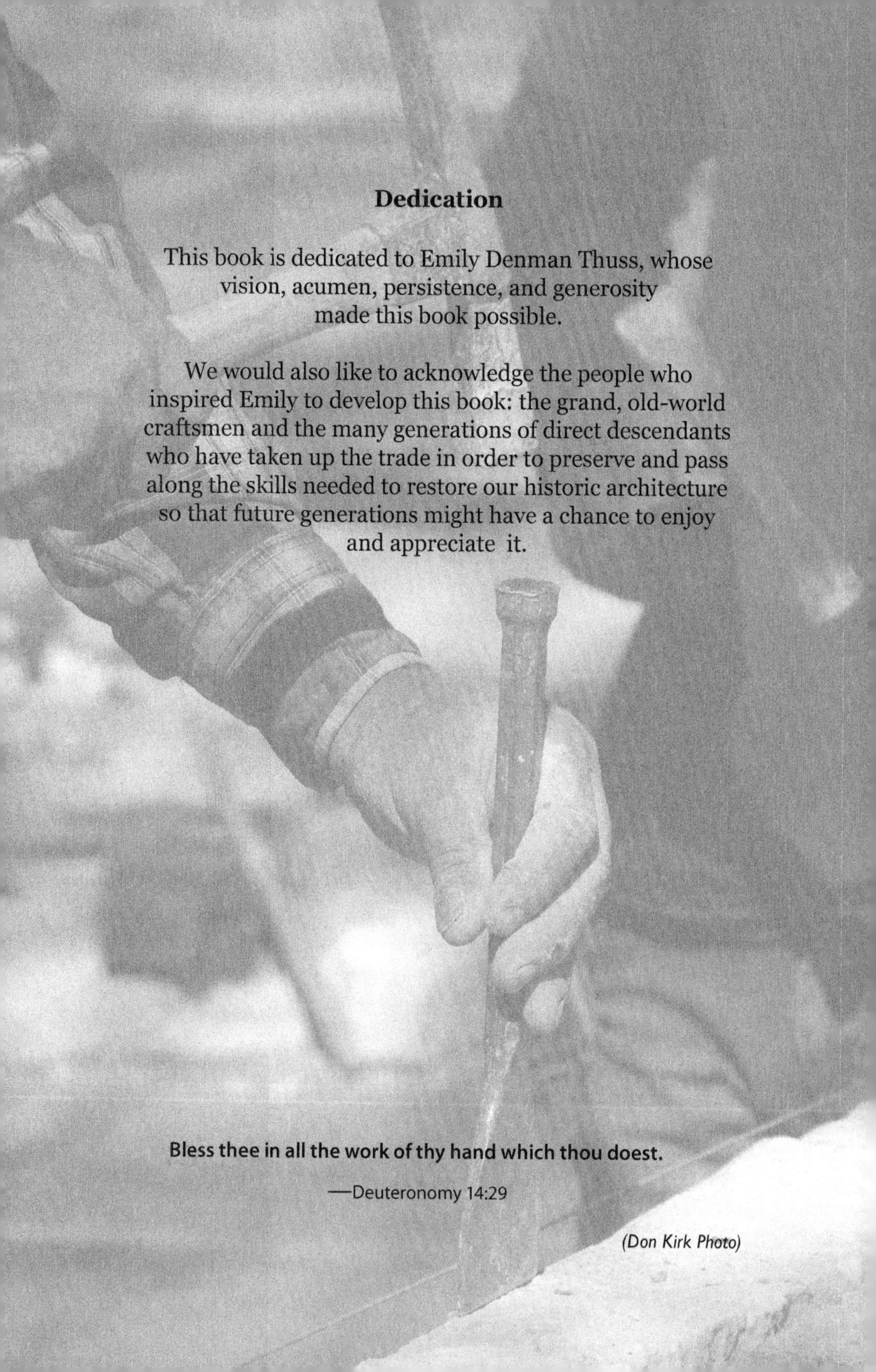

Bless thee in all the work of thy hand which thou doest.

—Deuteronomy 14:29

(Don Kirk Photo)

Contents

MISSION SAN ANTONIO DE VALERO

(Don Kirk Photos)

MISSION SAN JOSÉ

BEXAR COUNTY COURTHOUSE

The Building Arts Story

This book began with passions—passion for beauty, passion for the past, passion for quality and passion for the people who create and preserve timeless art for all to enjoy.

San Antonians and South Central Texans are blessed to live in an area that has, from its very beginnings, been filled with a wealth of beautiful buildings prized for their enduring beauty and historic value.

> *San Fernando Cathedral is the oldest continuously operating cathedral in the United States, first constructed in 1755.*
>
> *The Alamo, known worldwide as a shrine to the fight for freedom, was established as San Antonio de Valero. The extant mission church, or Alamo Shrine, was begun during the 1750s. Through the years it has been added to, changed, and enhanced—but always given loving care.*
>
> *The King William District is home to numerous houses of the rich and the humble, most dating from 1859, when the streets were first surveyed, until the last home was built in 1910.*
>
> *The Spanish Governor's Palace was built in 1749 and today is both a National and a Texas Historic Landmark enjoyed by visitors from all over the world.*

Missions Concepción, Espada, San Juan, San José, and their accompanying ranches and acequias, are part of the San Antonio Missions National Historical Park and remain active local Catholic parishes, which thousands enjoy each year.

Downtown has a host of other treasures, including the Southwest School of Art, established by French nuns in 1851 as the 'Ursuline Academy' for girls.

The Clifford Building, popularly known as the 'Royalty Coins Building,' was built in 1891.

The Empire, Majestic, and Alameda Theatres, built in 1913, 1929 and 1949, respectively, are now restored and being used as special venues. The Majestic and Empire showcase touring shows, the Symphony, and special events.

The 1929 thirty-one-story Smith-Young Tower (now the 'Tower Life Building') housed the first Sears, Roebuck and Company store, and today houses the offices of San Antonio attorneys and corporations.

The Building Arts Group and the San Antonio Conservation Society have teamed together to capture—before they are gone—the voices of the artists and craftsmen that help make possible the enjoyment of these poems in wood and stone by keeping them restored and conserved in all their glory.

The men and women in these pages are building arts practitioners, the creative souls who give us the plaster roses and wooden mantelpieces and iron hinges and hand-carved cabinets that make each building unique and noteworthy.

San Antonio is known for its ambience, its multicultural charm, and its preservation of historical spaces and places. But many of the building arts and the skills of the artisans who help make this city so wonderful are in danger of disappearing.

This book aims to preserve, at least in print and photographs, these traditional skills. In addition to the words and pictures, there are appendices listing projects by the featured craftsmen, and their contact information.

We hope you enjoy learning about these building arts and artisans and will appreciate—even more than before—how they have enriched our unique corner of the world.

—Barbara Dean Hendricks

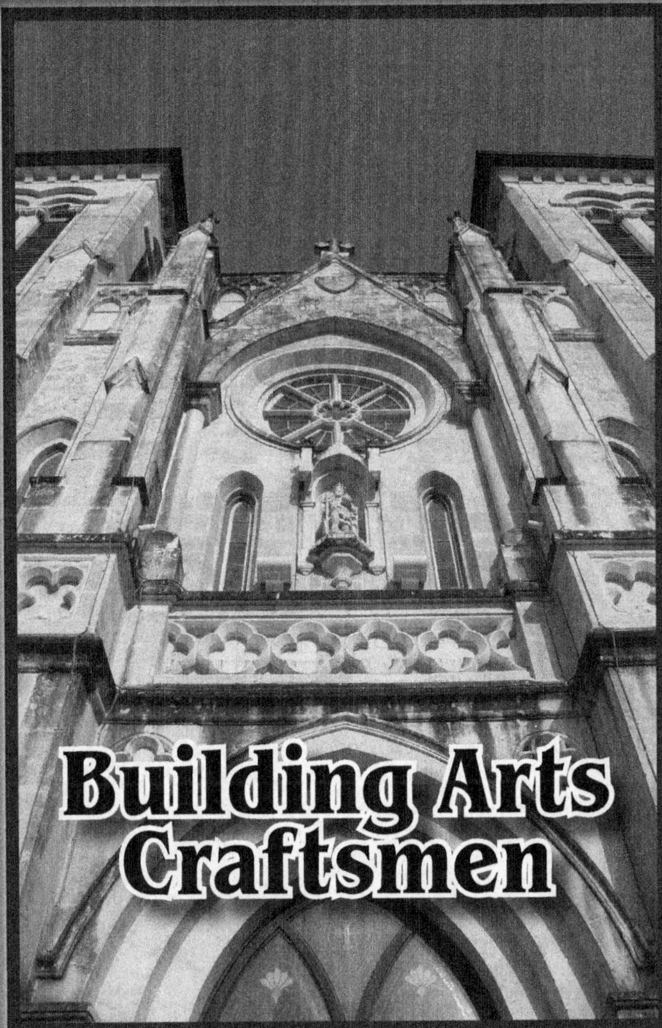

Building Arts Craftsmen

SAN FERNANDO CATHEDRAL

YMCA / ALDEN HOTEL

THE MAJESTIC THEATRE

DRURY

SAN ANTONIO, TEXAS
(Don Kirk Photo)

1. Johannes Scholze

(Monte Adams)

INTRODUCTION

JOHANNES SCHOLZE, woodworker extraordinaire, trained as an artist, studying art, woodcarving, and furniture design at a Dresden, Germany, art academy. Immigrating to San Antonio as a young man, his art—from mantelpieces to balustrades to the San Antonio city seal—graces such South Texas treasures as the Aurora, the University of the Incarnate Word, and City Hall.

When interviewed at the grand age of 102, Johannes remembered every piece he had carved in his eighty-year career. His tools are almost as old as he is, and he calls them by their German names even though he speaks mainly English and Spanish after settling in South Texas.

Working with various carpenters and cabinetmakers who followed his instructions and designs, Scholze did all the hand carving himself.

A list of his customers reads like a South Texas "Who's Who": Nixon, Dittmar, Kampmann, Steves, Frost, Brown, Denman, Fair, and even Lady Bird Johnson. Scholze carved the picture frame that holds LBJ's life-sized portrait in the Texas State Capitol.

(Monte Adams)

Johannes Scholze.

Johannes Scholze

WOODCARVER & FURNITURE MAKER

Johannes Scholze was born in Germany on October 23, 1901. His parents were both teachers, but at the suggestion of his uncle he attended the Dresden College Art Institute to learn woodcarving and furniture design.

Johannes told us that after the First World War: *Things were kind of tough over there and I decided to come to America.*

Johannes had relatives in San Antonio, Texas, and in 1925 he came to stay with them and look for work. He loved his adopted country and elected to spend the rest of his life in San Antonio. When asked about the city today versus 1925, he said, *Well, I liked it best when it was smaller. When I came here it was ninety thousand people. It was better then.*

Many long-timers tend to agree.

He was twenty-three-years old and single when he arrived in this country and in 1926 sent for his "schatzie," sweetheart and soon-to-be fiancée. In 1927, they were among the first couples married by a ship's captain in the newly inaugurated Houston Ship Channel.

Johannes and his wife had two sons, but lost his oldest son, Hans, when he was just 56. Their youngest son, Karl, married *Barbara Fortassain, a girl who has a history of French people.* The Scholzes had three grandchildren and three great-grandchildren, whom he adored.

Over the years, Johannes made regular trips back to Germany to visit his three sisters. His parents had died very young in their forties. Johannes and his wife were married for seventy-four years, until her death in 2001.

After all these years in San Antonio, Johannes said he still thought in both German and English and also spoke fluent Spanish.

At 102, Johannes was a living, breathing history lesson, a wealth of knowledge, and, as John Hall (another master craftsman) would say, he was *sharp as a tack.* Sadly, Johannes passed away peacefully a few years after our interview, but here is some of his story:

The Conversation

On a late spring day in 2003, we met with the 102-year-old Johannes at his rock home in northeast San Antonio. He swam every day in his backyard rock-edged swimming pool, rain or shine, and moved about the grounds using only a beautiful hand-carved cane.

> *An old friend from Germany had a cane like this. He was visiting here and I liked it. I told him, 'Let me copy it' and he said, 'Go ahead.'*

With Karl, his son, as interpreter (Johannes had become quite hard of hearing), we had a fascinating interview with the oldest living craftsman in San Antonio.

Training

> *I went to an old Academy in Dresden* [the College Art Institute, in Germany; it was called Kunsta Gava Shula] *and I took designing lessons. I had the first professor of Designing and Sculpturing, so he gave us instructions how to do it. You just had to go and do it, and then if there was something wrong with it, he would tell you how to straighten it. It took four years to graduate.*

Scholze's canvas roll of hand tools for relief carving and decorative work. (Monte Adams)

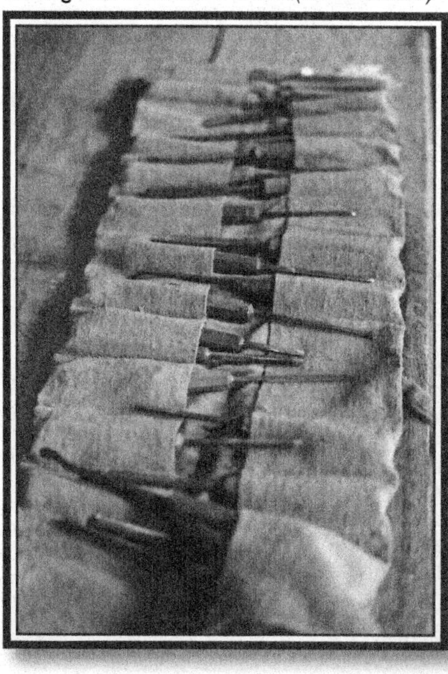

Johannes explained his decision to become a woodcarver, rather than enter another occupation.

> *I had an uncle who was an architect and he did designing and he suggested that I go into that school. That was my uncle, my mother's brother. He was a professor of architecture, a very smart man. He took me under his wing.*

Beginning in San Antonio

After Johannes arrived in San Antonio in 1925, he found work at the Prassel Sash and Door Company. It was a big plant, and the company worked on houses, churches, and large buildings. Some wood carving was needed so Johannes got the job, but after two or three months the company's need for a woodcarver came to an end and he returned to living with his aunt. Johannes was then hired at a number of other places, including the Alamo Planing Mill.

> *After I had been here almost a year, I asked my schatzie to come over here and she did, and we got married on the first of July aboard the arriving ship in Houston. It was quite a story because it was the first time marriage had been performed in the Ship Channel* [which had just been opened]. *It was a big story in the newspaper.*

Johannes Scholze at his workbench, carving on a plank of walnut. (Barbara Dean Hendricks)

Life for Germans in Early Twentieth-Century San Antonio

> *When I came here I didn't speak English, so I had to learn. There were two streets that were important in San Antonio, Houston and Commerce Street. There was a night school on Main Avenue, a vacant building, and these Jewish ladies had a club. The people who came from other countries and needed to learn English were taught there. It was free of charge and that's why I went, that's when I started learning English.*

He sought out German friends.

> *Well, I started to go to dances at the Hermann Sons Lodge and later on I went to Beethoven Hall (Maennerchor) and joined there.*

Johannes played piano accompaniment for the men's choir at Beethoven Hall for many years.

We had a group of young German people who all got acquainted with each other and decided to start a soccer club. We played soccer in San Pedro Park. That went on for several years. Later on, the group broke up and finished as if something had disappeared in the air.

The Depression

Things were hard for the Scholzes during the Depression—as they were for everyone—and Johannes moved his business several times, but his business continued to grow with the need for more cabinetmakers and men to do the painting.

I had cabinetmakers who helped me and I designed the furniture and the cabinetmakers assisted me in building them. I did the carving on the furniture, but the cabinetmakers did the

Johannes' woodcarving tools: chisels, gouges, skews, V-tools, knives, planes, dividers.

(Monte Adams)

> *cabinetwork. They did it according to my instructions and my designs.*

All the carving was done by hand.

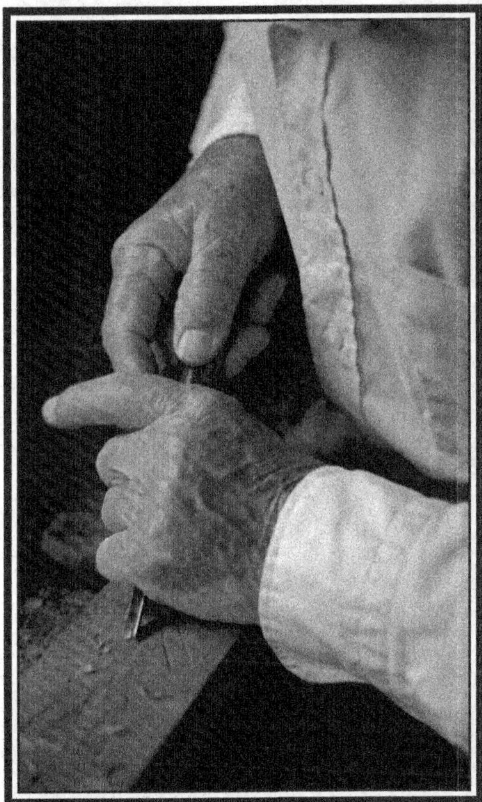

The strong, steady hands of Johannes as he demonstrates relief carving using a V-tool. (Barbara Dean Hendricks)

Carving Tools

> *I bought all these carving tools when I first started learning the trade. These tools are eighty years old, maybe even more. I had to add some new ones from time to time, but some of the originals are still here. They have different names, but I couldn't tell you the English names. I know the German names. For instance, in Germany and Europe, one tool that's kind of V-shaped is called 'gazfulz,' which means 'the foot of the goat.'*

Favorite Woods

Whenever he built furniture, Johannes always personally hand-selected all the hardwood he used.

> *The nicest wood that you can use for making furniture here in Texas is American walnut and oak. We made furniture with all kinds of wood—Honduras mahogany, Philippine mahogany—but my favorite wood is American walnut. It is beautiful wood and nice to work with—a nice grain and not too hard to work. Not as hard as oak, but it's just a lovely wood. In Germany, we worked with linden wood and walnut, even pine. Some things were made in pine, like buildings.*

Hand-carved fireplace mantelpiece by Johannes Scholze, Aurora Apartments, San Antonio, Texas. *(Monte Adams)*

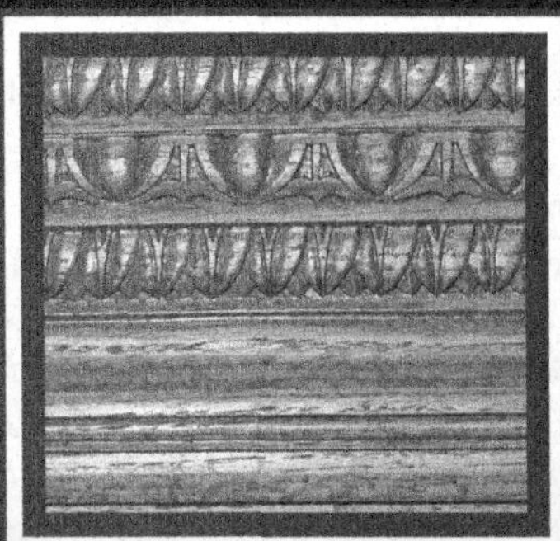

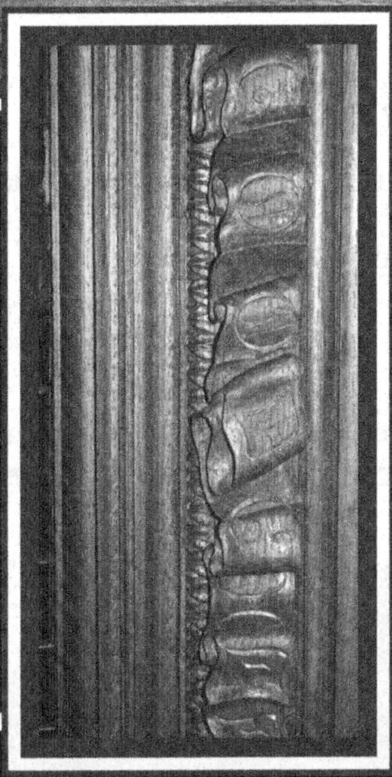

Fireplace mantelpiece details, Aurora Apartments.
(Monte Adams)

Old South Texas Families as Customers

Johannes' list of customers through the years reads like a history lesson.

> *Some of our very best customers were Mrs. Thomas Nixon and Mrs. William Taylor. I had the Dittmar family for customers and also the Kauffmans. The Steves family and the Frost family were good customers. Joe Frost and Mrs. Frost and her two children. Lutcher Brown. Yes, I made furniture for them too—and their daughter. Ralph Fair, Connie Fair. I did lots of things for Mrs. Fair out on the ranch* [Fair Oaks Ranch] *when they were out there. For Mrs. Lyndon Johnson, I made some things, but not directly for her. She had an interior decorator named William Sheffield and he did their decorating.*

Johannes carved the frame that holds LBJ's life-sized portrait in the capitol. He also did extensive carving for the Lutcher Brown family at their home on Ivy Lane.

> *We had the contract to do the millwork and then I did the carving for them. I made several mantelpieces like this one and the entrance and some gull wings that had some ornaments on it.*

Johannes said he was proudest of the entrance and some of the furniture at that house in Terrell Hills. For H.B. Zachry, Johannes made some hand-carved brackets for their house in Alamo Heights.

> *It had some beams on the ceiling and on the end of the beams there was a carving which I made and that's very beautiful still today.*

A stem-of-roses pattern on a chair back, carved by hand. (Barbara Dean Hendricks)

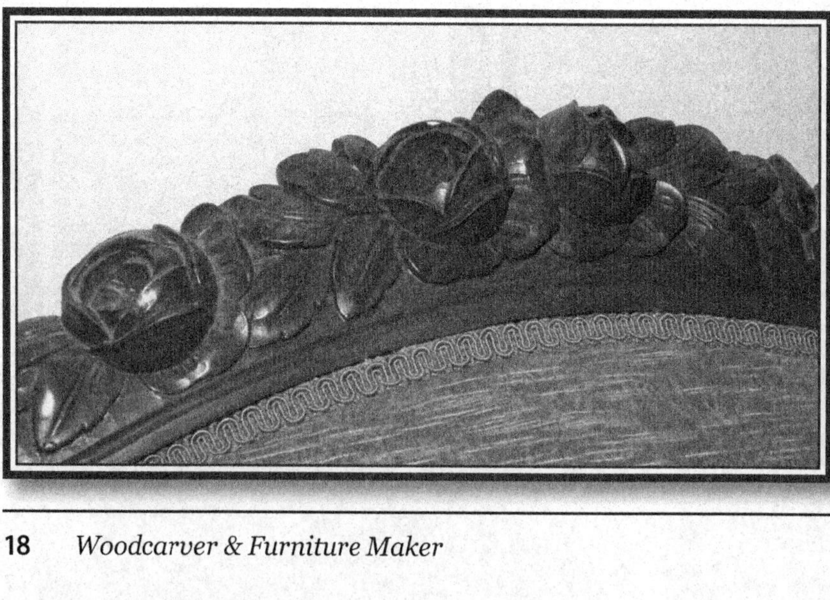

St. Dennis Chapel, O'Connor Ranch, Refugio County, Texas. Below, the chapel pulpit carved by Johannes Scholze. (Photos Courtesy Karl Scholze)

The O'Connor Ranch Chapel

One of his most interesting and magnificent assignments was for Mrs. O'Connor of the O'Connor Ranch in Refugio County.

I won't forget this one in Victoria. It was in the early 1940s and Mrs. O'Connor's son, Dennis, was a soldier overseas in the war. Mrs. O'Connor made a commitment that she would build a church if he came back in good health.

Her son did come back.

Harvey Smith was the chosen architect. He was well known for designing churches. He came to me to carve all the wood. We made most of the furniture and all the interior walls out of black walnut—quite a big job.

Carving detail on front of pulpit. Below, the baptismal font base, St. Dennis Chapel, Texas. (Barbara Dean Hendricks)

The contractor that built the chapel was Mitchell, and the contract was about fifty thousand dollars, a lot of money in those days. It took about three or four months to do the whole job. We took it down there and installed some of it—a circular stairway going up into the balcony. We made the stairway, too.

Mrs. O'Connor lived on the ranch and would come by to check on the progress as the church was built. It is still an active parish church today, still on the O'Connor Ranch, called St. Dennis Church, I believe, and can be visited by appointment.

The Ima Hogg Mirror

Like any good craftsman or designer, Johannes appreciated other beautiful designs and often made pieces for himself inspired by the splendid work he was asked to copy or repair. One example was a mirror in his home covered in gold leaf.

We copied it from a mirror that Miss Ima Hogg had. This is more beautiful than hers. Hers was getting dark and it was very old. It was probably two or three hundred years old. We made this about fifty years ago. The gold leaf was put on by a German painter by the name of Hillenbach.

General McDermott's Desk

In 1974, working as a subcontractor to Prassel Sash and Door Company, Johannes made a replica of a museum-quality piece for General McDermott's new office at USAA headquarters in San Antonio.

General McDermott sitting at his custom-made Mahogany office desk at USAA headquarters.

Johannes' carved signature on the underside of the General's desk. He crafted the desk in 1974.

Furniture leg and foot detail. The desk was an exact copy of President Jefferson Davis' personal desk.

(Monte Adams Photos)

(Don Kirk)

Front door to the Spanish Governor's Palace, San Antonio, Texas.

General McDermott had seen a beautiful desk in Colorado Springs at Colorado College. The college director had the original desk of Jefferson Davis. They took me to Colorado, I took pictures, and we made the desk. The wood is Honduras Mahogany. I was already getting old, but we still made it.

The San Antonio Coat of Arms

Johannes' first cousin, once removed (his mother's first cousin), was an engraver and made designs in metal—like badges for the San Antonio police. Together they had the idea to design a coat of arms for the City of San Antonio. This opportunity came when Johannes had only been in the United States for a few months.

I carved the coat of arms in wood. We made it and gave it to them and they accepted it. They did not order it, it was a voluntary thing—a gift.

Original coat of arms carved in wood by Johannes Scholze in 1925. Later, a larger version would be molded in plaster by Battersby. (Monte Adams)

The design was created to relate to the history of Texas.

For instance, the Alamo was something that everybody knew what it stood for. 'Libertatis Cunabula' is the Latin name. It means Cradle of Liberty.

They presented it to Mayor Tobin at City Hall, sometime in September of 1925. The design of the coat of arms was used to create the city seal, found today on stationery, blueprints, and all the city's business cards.

Important San Antonio Buildings

Johannes worked on the original Motherhouse at the University of the Incarnate Word and the Milam Building. He also made furniture for Mission San Antonio de Valero (the Alamo).

I made a conference table and about a dozen chairs [for the Long Barracks]. *I did some carvings on the door to the Spanish*

Johannes Scholze, San Antonio, Texas.

(Monte Adams)

Governor's Palace and all of the furniture for Fort Sam Houston Officers Club through the decorator for Joske's. I had two upholsterers working from their home doing all the upholstery pieces, so it was quite busy. That was around 1936 or '37, something like that.

Johannes' handiwork on his own front door. (Monte Adams)

His Home

Johannes built his own home in the late 1940s and told us the story.

I did a lot of work for an architect and he used to come to my shop. He caught me there one time sitting at a drawing board and I was sketching a house and I told him I wanted to go ahead and build myself a house.

He looked at it and he said, 'Well, you should not do this. This is a professional job and you'll have too many mistakes in there if you design it.' I said, 'Well, what are we going to do about it?' He said, 'Well, I may design it for you and only charge you for the actual cost in my office,' and so I said, 'Okay, go ahead.' So he came out here and looked at the lot and then he designed the house and that was all there was to it.

In those days you could go to a lumber yard and buy plans for a house for about $25 or $30, but when [the architect] *got through with the plans for my house, it was $225, and I said, 'Oh my God!' I had a good Mexican carpenter who could 'see' the plan and he knew just exactly what needed to be done. We started on May 1, 1947, and we moved in on November 15, 1947.*

"This is so different now, what is being done. A lot of woodworking is done now by machine and it's a changing world." —Johannes Scholze

Newel post detail.

Staircase balustrade,
Col. George W. Brackenridge Villa,
University of the Incarnate Word.

(Monte Adams)

Johannes' house was graced by many examples of his finest wood carvings—from architectural touches to museum-quality furniture. The mantelpiece was the beautiful focal point of the living room.

Apprentices—A Lost Art

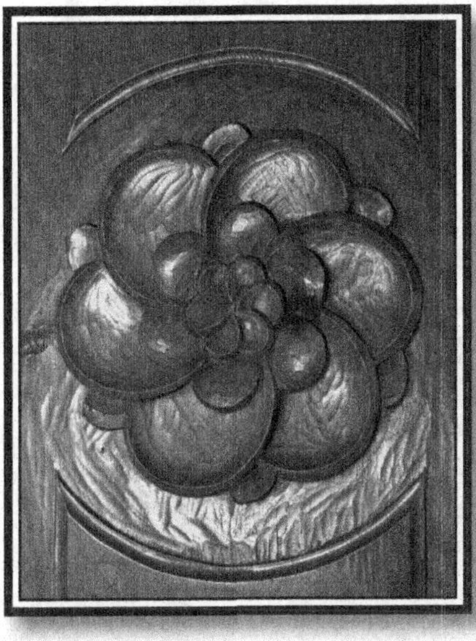

Hand-carved rose. (Monte Adams)

Johannes never had any apprentices. He said that in this country, apprenticing was "out of style," but a few people did come to learn from him through the years.

> *I had some people, some ladies, who wanted to do some carving and I gave them a few instructions, but not any real thing. I mean, nobody ever took up carving like that. I had some woodcarvers from México at one time. They were good men.*

Johannes did not feel that woodcarving was a lost art just yet.

> *No, no, it's still going on in Europe and all over the world. The man that is renting our shop now, he is importing furniture from India and there's a lot of carving on that too. It's a little bit crude, but still it's not bad. There are some woodcarvers somewhere in the United States, but not in the South very much. The industry seems to be in states like Wisconsin and California. I'm sure there are some people who do good woodcarvings, but, this is so different now, what is being done. A lot of woodcarving is done by machine now and it's a changing world.*
>
> *There used to be these planing mills in San Antonio—four large planing mills, in fact. There's not even one anymore. Now buildings are built out of steel. The materials come from a factory. Planing mills like we used to have had hundreds of workers.*
>
> *Some things cannot be made in factories—like paintings or even woodcarvings. Few can say 'I want to do this' and make a living out of it.*

The result is less handwork, less craft work, and a great loss.

An elaborate, highly detailed, beamed ceiling at the home of H.B. Zachry.

The beams were finished out with brackets carved by Johannes.

(Don Kirk Photos)

Johannes Scholze at his home in San Antonio, 2003, 102 years young. *(Barbara Dean Hendricks)*

Pulpit and stair balusters carved by Scholze, St. Dennis Chapel, O'Connor Ranch, Texas. *(Courtesy Karl Scholze)*

His Legacy

When asked what he most wanted people to remember about him, Johannes reflected and replied.

> *I wouldn't know what to tell you. All I can say is I enjoyed my youth in Texas. I have been here seventy-eight years and it has been very nice, very good. I am happy here. Even though we went to Germany to visit, but it was never the same again; so I became a native here. I've been very happy with it* [the craft of woodworking] *and I'm still part of it now.*

Until his death, Johannes Scholze was the very definition of a master craftsman. His work remains an ongoing inspiration.

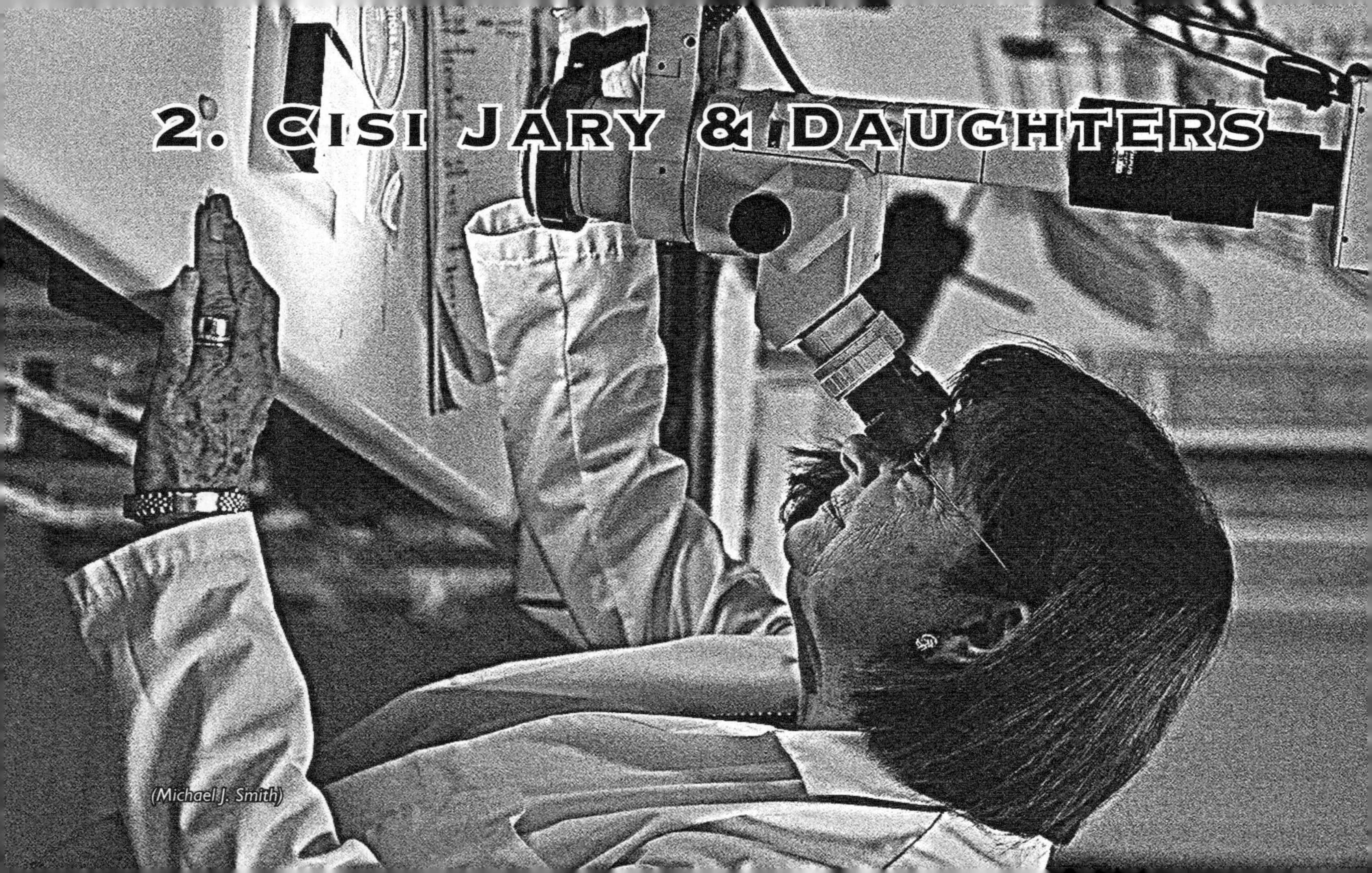

2. Cisi Jary & Daughters

(Michael J. Smith)

INTRODUCTION

CISI JARY, owner of Restoration Associates Limited, began business with the restoration of an opera house in Galveston, Texas in 1985. This soon led to a unique family business. Today, along with her two daughters, Elisa Jary Nieto, and Pamela Jary Rosser, the company is known around the country as the expert to hire for the restoration of historic paint. Cisi, who frequently visited her preservation-architect husband's work sites, noticed that general paint contractors were doing the wall art—from gold leaf to ornamental plaster—and it wasn't working.

Cisi's background in art, art history, and the history of architecture led her to believe she could save the historic paint and bring the decorative art back to its original glory by applying some of the same methodologies used to save and restore mural art and works on canvas. She likens the process to paint archaeology—removing the paint, layer by layer, using strong toxic chemicals, dating the layers chronologically, researching the chemical composition of the original paint, and recreating and applying paint as close to the original as possible.

Cisi tells the artists working for her that no artistic license is allowed. The work belongs to the original artist, and their job is to preserve that work. With projects like the Aztec, the Alamo, and the missions behind them, Cisi knows there is always more to do: "It is so important for young people to know their heritage".

(Michael J. Smith)

Pamela Jary Rosser, Cisi Jary, and Elisa Jary Nieto.

Cisi Jary & Daughters

Paint & Finish Restoration Artists

Cisi Jary, owner, Restoration Associates Limited, has—like most of the artisans featured herein—an interesting "back story" about the beginnings of her life's work and the establishment of her business. And like so many, it relates to family.

Cisi relates,

> *My husband, Lloyd, is an architect here in town and he does preservation architecture. I would sometimes go with him to construction sites where they were doing preservation and restoration work, and they were letting the general painting contractor do the wall art! The gold leaf, and whatever, on ornamental plaster. From my background in art, art history, and the history of architecture, I knew that those paintings on the walls were decorative works of art. It was sometimes done by itinerant painters who did this for a living so they could afford to paint fine art. I believed that this historic paint could be saved if the restorers would apply some of the same methodology they did to mural art and works on canvas.*

The First Job

In 1984, Cisi bid on her first job. There had already been two contractors on the project. She remembers:

> *There had been a hurricane in the '70s. There wasn't very much historic fabric left in the building when we got there. We didn't do any paint analysis because there was none of the historic plaster left. So, we recreated a lot of it from photographs.*

The project took about four months and included a breezeway area around the room and ornaments on the ceiling and around the proscenium. The entire job was completed by only three women.

> *It just worked out that way.*

Cisi Jary has had projects that included one or two men who were artists, but the main difficulty is height.

> *Most women are 5'-4" or 5'-5," somewhere in there. If a man works for us and he is six-foot tall, the scaffolds are not the right height for him, and it is uncomfortable. It's either uncomfortable for him to work with us or it's uncomfortable for us to work with him. So, it just has happened that mostly women have worked our projects.*

Her Career Preparation

Jary's educational background gave her a unique preparation for this highly specialized career. She was trained as a fine arts painter. She believes both Art History and the History of Architecture courses were the most useful to her.

> *I think they are really strong requirements for success in the business. I had one professor from the University of Texas in Austin who was from England, and in order to take his class and survive you had to learn all the traditional classical painting methods. You had to prepare your own canvas, make your own gesso, and so I had that background already. As far as casting ornamental plaster, we didn't make that many molds in sculpture class.*

> *I started in Florida—Rollins College in Winter Park—and then transferred to the University of Texas at Austin. Mostly because my Dad told me if I didn't go to school at UT, I needed to go get a job; he wasn't going to pay out-of-state tuition.*

Cisi was happy with the result.

> *I had no idea that they had a very big Arts Department and I found it quite wonderful.*

She married before she graduated and moved to San Antonio. She recalls,

> *I went back to Trinity and took a painting course. Then, after we had our fourth child and after she went to first grade, I went back to school. This time I went to Incarnate Word and took painting. The professor was a wonderful draftsman; you learned to do a very detailed charcoal drawing before you ever painted. You had to know draftsmanship, which was difficult for some of the students who didn't have the background.*

A Family Business

Jary and her two daughters, Pamela Rosser and Elisa Nieto, are the principals of Restoration Associates Limited.

Both of my daughters have the same passion for restoration as I do. And they know the same procedures and conservation methods as myself. Lloyd and I are very fortunate that they have the same passion for historic architecture as we both do. And it's very interesting to see their approach to things and see how they handle the clients.

Applying gold leaf to a plaster casting at the Empire Theatre. (Tommy Adams)

At the time of this interview, Elisa was working on the Sutton County Courthouse in Sonora, and Pamela had just finished the Aztec Theatre and was preparing to start restoration work on two courtrooms in the Bexar County Courthouse.

How the Family Got Involved

Pamela, the youngest daughter, was a senior at Churchill High School when Cisi's conservation career began. Pamela attended UT for one summer and then came back to Incarnate Word.

> *While Pamela studied abroad in Italy, I gave her the name of some conservator friends of mine and she looked them up in Florence and ever since then she has been very, very interested. Elisa is the oldest. She studied Architecture at A&M and so has the architecture background. Pamela has more Art History, History of Architecture and Applied Arts than Lisa, but the main thing is that they have passion for restoration work. If you don't have passion for your work, I don't know how people get up every day and face it.*

Restoration Associates has stayed busy since that first job at the opera house—for twenty straight years! Though the company has a website and has advertised in a few professional periodicals, excellent work and word of mouth have been their main promotional tools. In addition to having her daughters working side by side with her, Cisi and her husband have sometimes worked on the same projects: St. Mary's Catholic Church in downtown San Antonio, a Historic Structure Report on a home in New Orleans, and the Magnolia House on the fairgrounds in Dallas are just a few examples. Cisi did paint research for him on the Library at Keystone School as well. Cisi Jary's enthusiasm for restoration rubbed off on her husband Lloyd.

> *When he first started his practice, restoration wasn't very popular, but the state of the art has really changed. Now people are interested in restoring their buildings, or at least in finding out what the historic paint was in the structure.*

Interior archway detail, St. Mary's Catholic Church, San Antonio, Texas. (Don Kirk)

Preservation architecture has special parameters.

> *To be a design architect is relatively simple. To be a preservation architect is difficult. They already have the building, but they have to add everything that brings it up to code within that same space. And with that, the owners all want new amenities. The process then becomes much more difficult and challenging.*

The Challenges

Cisi and her daughters do more than supervise. Their handprints, so to speak, are everywhere. It's important to have good light, Cisi notes, and

> *we've gotten very particular about the height of our scaffolding. If you have it set right you work kind of out here in front. The only time it's really tiring for all of us is when there's something like a vault—a decoratively painted vault door—in a courthouse that cannot be removed. On the restoration of the very lowest part of the door, which is four inches from the floor, the only way you can do it, is to sit or lie on the floor.*

The Artist's Tools

When questioned about tools for her craft, her first response of course is brushes, good brushes.

> *Mostly all have to be natural hair brushes—bristle brushes, sable hair brushes. Specialized brushes are used to apply gold leaf so that you get a very even layer. We do work in water-based paint and that also requires very good brushes. The problem in restoration is if you don't follow the very same steps that the original artists did when they painted, it will never look the same.*
>
> *Microscopes are used to analyze paint. They tell us if there is a varnish layer and how many paint layers there are. Usually between each paint layer there will be a very thin layer of dirt and that is how you distinguish one layer from the next. Colors don't really fade that much over time if you harvest the samples from an area where it wasn't exposed to much light. They do oxidize if you're mixing oil with yellow, though. You have to take all of those things into consideration, but also you have to consider the patina of the historic building itself.*

Paint Archeology

Removing paint, layer by layer, is a very delicate and difficult operation.

> *We use a very toxic chemical. You want to be able to take it down layer by layer so you can set up a chronological dating as to when*

Tools of the trade. (Michael J. Smith)

the paint was applied and how long it had been there. Removing layers of paint is a paint archeology.

The toxic chemical we use is called Metholine Chloride, a gel. There are some others that are not quite so toxic, but you want to be able to take it down layer by layer. There are other paint strippers that work, but they don't go layer by layer. Potassium hydroxide takes it all off, but then you can't analyze it.

Dealing with Minerals, Chemicals

If we find decorative painting that has lead, we encapsulate it. It's just like putting an isolating layer of varnish and as long as you don't disturb it, it's perfectly safe.

For protection, her daughters always use a mask and gloves. Cisi says that she routinely wears gloves, but

sometimes I don't use a mask. I get claustrophobic wearing a mask. They are carbon filter masks. Sometimes, if we're consolidating limestone, say with ethyl silica, we have to put on those hot white suits that are very uncomfortable. We only do it in wintertime.

St. Joseph Catholic Church. (Don Kirk)

The Process: Creating Paper

Paper assumes an intricate role on many jobs.

Using an adhesive and several layers of special papers that we ordered from California, we had several papers to test. They were all like fiber paper but a bit thicker.

The project was St. Joseph Catholic Church downtown,

the one everybody refers to as St. Joske's. When we did the job, the ceiling panels over the main altar were

The interior of St. Joseph Catholic Church, completed in 1871, San Antonio, Texas. Below: The church became known as St. Joske's in the 1940s after Joske Brothers expanded around the historic Gothic-style church when the parish refused to sell. Only the facade of Joske's is now being saved, with a total reconstruction of the interior for retail shopping in 2014 and 2015. (Don Kirk)

> *falling and from twenty-five feet down it looked like they had been painted on canvas. But, when we got the scaffolding up and started our conservation treatment, it turned out that all that ornamental painting was on cardboard! The whole ceiling is cardboard and the roof is just a metal roof. There's no insulation and just this layer of cardboard panels. The original church dates back to about 1860, but this particular decorative scheme that's in there now dates back to the late '30s or the early '40s, I think. When we got up there we discovered that we had to treat it as a work on paper instead of a work on canvas. So, in order to repair the tears, we had to literally make paper because some of it was missing.*
>
> *We had to make it on site up there on the scaffold. And we used special papers that we laminated together until we brought it up to the same level as the cardboard and then we sealed it and painted it. It was painted with oil paint, so we put a protective layer on it, and it had gold leaf on it.*

The Process: Japanese Fiber Paper

Jary explained another tedious, but often necessary, process in restoration.

> *You figure out how long it's going to take to remove that layer and you do it in time intervals. We like to use Japanese fiber paper a lot as a facing. First, you remove the paint layers. Then in order to reattach the paint, you put a facing (the Japanese fiber paper) and inject the adhesive behind it. That way nothing falls away. In some cases, you apply that first with mineral spirits and then you can put on the stripper. The paper protects, so you don't lose any of the decorative painting underneath.*

Examples: The Missions and the Alamo

> *We have flaking paint at the Alamo and all the missions. They were frescoes painted over with white wash—many, many layers in some cases.*

Cisi learned this process when she worked at Mission Concepción with the Conservator Paul Schwartzbaum from the Guggenheim.

> *Paul is Chief Conservator in Venice and in New York, and he was our team leader. Anthony Crosby was in charge, and he and Schwartzbaum taught those of us who were actually doing 'reattachment,' as it's called. This is a common practice on frescoes and on paintings done on canvas when you have to reattach the paint layer.*

Along with Guido Bocacelli from Florence, Jary worked on the well-known so-called "Eye of God" room at Mission Concepción.

> *We started chipping away Portland cement patches that were on top of part of that face of the sun. There was another eye and a mustache and a goatee, so all of a sudden 'God' took on a very Spanish flare and had a second eye so the room couldn't be called the 'Eye of God' room anymore! So, now they call it 'The Library'!'*

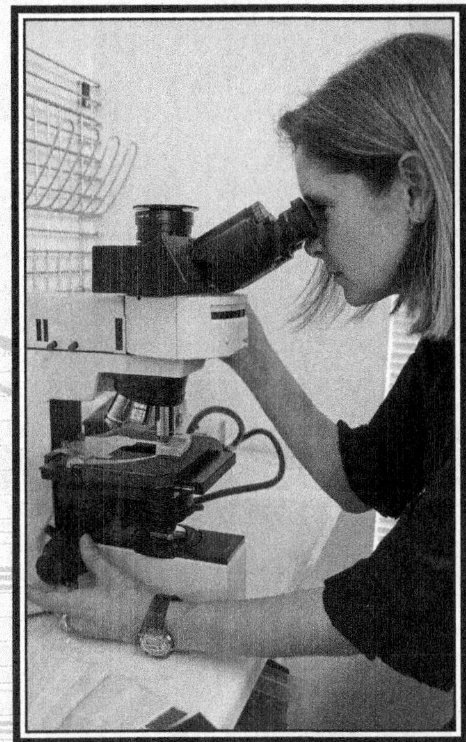

Pamela Rosser analyzing paint samples. (Michael J. Smith)

Cisi's favorite Spanish mission is Concepción.

> *Mostly, because there you have most of the original decorative painting intact. At Mission San José you only have it in the arches in the Convento area. At Mission San Juan we worked in the Convento looking for decorative painting. `But it was all just white plaster. We know there's decorative painting in the church at Concepción, because if the light is right and you have track lighting like you have there, you can see where the pattern was incised into the plaster. But it's been plastered over and painted so many times that I don't know if they will ever retrieve it. It would be a wonderful thing if they could ever raise the funds to do it.*

> *If you look at the façade at Mission Concepción after it has rained, the rain brings out the color. You can see pigments still up on the walls on the church, so it's still there. There's still red pigment on the bell towers, too.*

The conservators did uncover ornamentation—the design that was the stencil pattern, a flower and a pomegranate.

(Don Kirk Photos)

Original seventeenth-century, and later, fresco-seccos made with tinted lime washes were restored by Cisi and her daughters in the Convento of Mission Concepción. Christian, Spanish, and Native art symbols were found under hundreds of years of dirt and grime.

Top Left: Colorful decoration boldly surrounds the Convento doorways.

Bottom Left: In the sacristy, a fresco that looks like a draped curtain.

Mission Concepción, San Antonio, Texas

(Don Kirk Photos)

Above: A fresco with cross and potted plants over a door.

Below: On the ceiling, the formerly "Eye of God" image now has two eyes after restoration.

An eight-pointed "star" pattern on the ceiling of the Convento. In color, it would have red ochre, yellow ochre, and black and white elements.

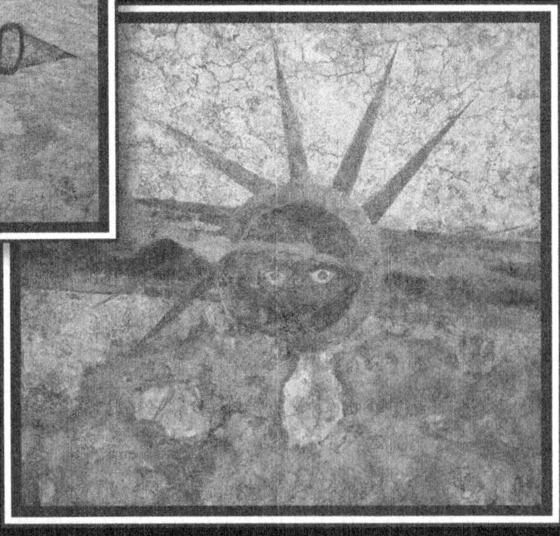

Restoration Documents for Mission San José, 2010. (Courtesy Rubiola Construction)

International Connections

There might be a connection to a craft school in México City.

> *There was a painter, Miguel Cabrera, who taught there at the time the missions were being built. And in his paintings you could see that same style of pomegranate. We also found the same little elements on a smaller scale in the sacristy at Mission Concepción.*

Cisi notes that the famous Rose Window at Mission San José includes a pomegranate element, but most experts don't think that the designer is the same person that decorated the missions.

> *In the sacristy that we worked in a year ago last February, it's primitively done. The pattern that we found around the barrel ceiling has a flower, a four-petal flower that's inside a triangle. On the front of the church carved into the limestone is a diamond with the same four petal flower. What we think—and we only found a few fragments of pigment—was that it was a mural design. The part of the triangle we uncovered, had the other part on the other side, and it would create a diamond, which was also on the front of the church.*

> *Originally all the churches were plaster, because that would have been a protective coating on the limestone. They have*

One of the interior elevation drawings showing the areas that need plaster restoration, Mission San José. (Courtesy Rubiola Construction)

photographs of Mission San Juan from the 1920s, I believe, that show ornaments on the wall. I think all of the churches were decoratively painted because that was just the way they did the churches. I'm sure if it's in Concepción and San José, it's bound to be in the other two mission churches as well.

Another Mission Tale—Plaster and Weather

We did an assessment of Mission San Xavier del Bac, near Tucson, Arizona. Most of the church had already been restored. Paul, the colleague that I mentioned earlier, and two of his teammates from Italy came every February for five years and restored the church, and it is magnificent! Located outside Tucson, it looks like a big white dove in the middle of the desert because it's bright white. It was not painted on the outside that we know of, but they've replaced the plaster. When they take out the plaster, there's no historic fabric left and I don't know if anyone has ever tried to document it.

They re-plastered it about ten or fifteen years ago and it was the wrong mix. They had to take it all out and replace it with plaster that was compatible with the original, which was lime and cactus

Bell Tower of Mission San Xavier del Bac
Located a few miles south of Tucson, Arizona. It was re-plastered with the wrong material and had to be removed.

(Don Kirk)

juice and sand. Portland cement plaster is terrible; it's too stiff. Lime plaster will move and doesn't expand and contract at the same rate as Portland cement. It harbors moisture, and that's the enemy of everything we do.

The Process: Pigment Reintegration

One of the things that's very important in conservation is when you're doing pigment reintegration. Whatever field you're working on has to be a shade or two lighter than the historic paint so that a practiced eye can look at it and see exactly where the reintegration occurred. The average viewer is usually seeing it from at least twenty-five feet away so the average person's eye will make the adjustment and not notice it.

Mixing Paints

On the missions, we a used powder pigment from Europe. We don't use any tubes or anything like that. We mix the powdered pigment with water and a little PVA (polyvinyl acetate). Sometimes we don't use PVA; we will use skim milk as the binder. In fact, most of the time that's what we use—skim milk. When you slice lime, there's a water residue at the top of the lime. You take the water and add skim milk and add powder pigment and there you have it.

Naturally Based Pigments

Cisi starts explaining pigments that come from nature.

Cobalt blue is one. And all the cadmiums are very expensive. Cadmium is from the earth and so is cobalt. It's a rock. Real beautiful blues are from lapis, the actual stone. There's a book about Michelangelo's painting the ceiling of the Sistine Chapel. It's fiction, but it's really interesting because it tells you what part of Italy some of these pigments come from. Like sienna is from sand. If you go to George West, south of San Antonio, you can see red dirt. The sand around there, the earth, is deep red. Most of the time they didn't use those colors on the walls. If you see cadmium, they imported it from Europe.

Colors at the Alamo

Cisi explains how her curiosity led to new discoveries.

We took some paint samples because we wanted to know where the red came from. In paint analysis we can do some of it here with our equipment, but there are much stronger microscopes that can analyze the pigment particle itself and that helps date it and determine the origin of the paint.

The paint analysis company identified the pigment and determined that in circa 1729, the Alamo pigment would have been imported from Europe.

Then we found that there were craft schools in México City. When they were tearing down all the Indians' Aztec temples and building Catholic cathedrals they had to train people to do this. And when they did all the building in México City, they ran out of trained craftsmen. They taught the Spaniards first, then they taught the Spaniards who were born in México, and then the Creoles. When they ran out of those people, they started teaching the Aztecs or the Native Americans—whatever tribe they thought showed some skills in painting.

Of course, the Indians already had skills and some methodologies better than those from Europe, but in order to imitate the Renaissance style, the Europeans had to teach them.

Summarizing, she concludes:

What we found at the Alamo, that bright red was actually imported and probably came from México. We haven't been able to trace how it came up here, but we know it was through México City. Everything shipped from Europe went to México City. And then it was shipped north because they painted all their churches.

We know the Alamo was painted. We also think that in the sacristy where we found the ornamentation there were five different designs—stencil patterns—that go around the walls following the barrel ceiling. And, we're pretty sure that all that was there during the battle.

The reason this was not known before, she conjectures, is because

it had been painted over, of course, with white wash. When the quartermaster occupied that room they cleaned it up. It had smoke from the fire and everything. So, all that was left is up high where nobody could reach to carve graffiti into it or destroy it some way.

Colors at the Missions

She tells another "investigative" story.

When we were doing some assessment at the Convento at Mission San José, some of the original stones in the wall were red sandstone. Piedras Creek, which is south of town, and also Atascosa County, have a lot of this deep red sandstone.

Original eighteenth-century frescoes on the walls of the sacristy of Mission San Antonio de Valero (the Alamo) were discovered by accident under layers of grime and whitewash. The preserving whitewash was probably put there in the 1840s when the chapel was used as a supply depot. *(Courtesy Restoration Associates)*

So, we gathered some up and we soaked it for about two weeks and then ground it up with a mortar and pestle. Then, we mixed it with skim milk and painted it on some plaster samples that we had cast specifically for this, just to see what would happen. It created a kind of burnt sienna—a brownish red color—and it's permanent. It doesn't rub off or anything, because milk is the binder. We thought perhaps when they were building the missions they had to use materials that were readily available. Red sandstone would have been available and could have been one of the colors.

The Native Americans' Use of Color

Cisi explains the sources of our colored pigments—like red from iron oxide, black from charcoal, or yellow from peroxide.

One of the things that we are always asked in workshops is how the Indians used the dried bodies of female cochineals, that little insect that lives and lays its eggs on cactus. Do we ever find a pigment with those particular colors all through it? It's crimson-colored and it's a permanent dye. You can tell if you ever get it on your fingers.

Lots of times yellow oxide or yellow unguent was taken from different sands or clays from the bottom of rivers and streams. Onion also makes yellow ochre. From what we have analyzed, it has not been from different sands, so I don't know where they could have gotten them except from onions.

To get a yellow like that on textiles, they use the cochineal. They used the cochineal to stain their deer hides, and weavings, and

The Jarys performed a conditions assessment on eight murals in the Navajo Nation Council Chambers at Window Rock, Arizona. The artist Gerald Nailer Sr. painted them in 1942 to depict the history of the Navajo People. (© The Trustees of Columbia University, Media Center for Art History, Department of Art History and Archaeology, Photograph by Gabriel Rodriguez.)

things like that. But, they also had a bright red and that's one thing I wanted to research. The Native Americans used cadmium red. Where did they get that color?

Interestingly, we did the assessment for the Navajo Nation Council Chambers in Window Rock. There are eight murals all the way around the tribal room. They were all painted with casein in the 1950s. Casein was before acrylic was really popular. You could buy it in tubes and that's what they used on Portland cement plaster.

Cisi worries that the paintings may not last.

It needs work. When they get snows, some water leaks drain down on maybe three of them. And the plaster is crumbling over the entry doorway. They haven't done anything so far.

They want to update it, but 'updating' to them is getting all the Elders to have a computer on their desks. That's what they want first rather than save their heritage! I believe they just don't think the paintings will fade away. It's very sad.

Materials: Prickly Pear

At Mission San Xavier del Bac, they analyzed the plaster and found that it had protein and carbohydrates in it. This suggests that the old traditional Native American plaster also had this in it.

So, they replaced the Portland cement that they had previously used with plaster that was made with boiled prickly pear as an additive. I don't remember the architect, but there are photographs of that procedure. And, if you see it, it's very white and not as hard as Portland cement plaster. I've been out on the roof there and lime plaster is softer and you can stick your fingernail in it.

Portland lasts longer than lime and, when it was invented, Portland became the answer to everything—they use it everywhere. They both last, but Portland cement harbors moisture. So, if you have paintings on it, they're eventually going to have problems because of the moisture that you get from rising dampness—which is very bad in San Antonio and around this area. In New Mexico, they do have rising damp, so to speak, but not like we do. New Mexico is much drier, so even some of their very old adobe is still standing. Whereas here, it's crumbled and falling away.

Building to Last

I'd like to ask some of the architects, when I have an opportunity, about building buildings to last. The same materials are available to architects year after year after year, that's why they design the way they do, because of what's available, what's economical.

Frank Lloyd Wright started doing housing with low ceilings. He was the one who revolutionized home building. Prior to that, you had tall ceilings. Then with all the air conditioning equipment and lowering the ceiling another foot and a half, you cut the expense.

It's just part of evolution. After WWII everything had to be slick and clean. They wanted everything to look modern. And now they're removing those facades. Look at downtown Ft. Worth, if you've been to Ft. Worth in the last fifteen years, downtown looks fabulous with all the restored buildings. It's fun to go there.

Now, I think we're coming back to building rooms that have higher ceilings, but not with the same materials as in the past. A lot of buildings these days don't have any wood in them—it's polyurethane. It looks like wood, but it's not. It will last, but it won't be as pretty since it won't age to a beautiful patina.

A Mayan art theme at the Aztec Theatre, San Antonio. *(Courtesy Restoration Associates)*

Large, colorfully painted Mayan ornamental plaster in the Aztec Theatre, 2005. *(Tommy Adams)*

Detail of a hand-painted Mayan plaster band in the Aztec Theatre. *(Tommy Adams)*

Fort Sam Theatre. Original stencils on beams in entrance lobby and in 2nd-floor lobby. Some were completely painted over. All beams received conservation treatments and painting. *(Courtesy Restoration Associates)*

Fort Sam Historic Theatre. Murals and original finishes were recreated for the entire theatre. Two 28-foot tall murals flank the stage opening. Paint analysis was used to identify the original color palette. *(Courtesy Restoration Associates)*

Briscoe Western Art Museum entrance lobby during restoration. The historic plaster detailing was originally faux grained to match the original wood panels. *(Courtesy Restoration Associates)*

Before and after shots of a star detail on a ceiling panel. *(Courtesy Restoration Associates)*

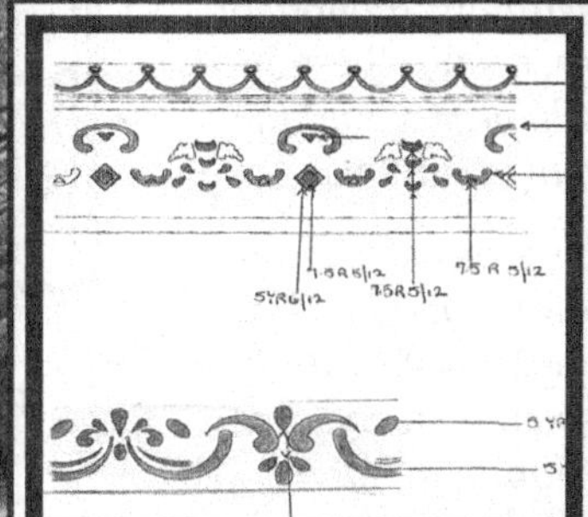

Sketch indicating paint colors to be used on a fresco. *(Courtesy Restoration Associates)*

Fully restored entrance lobby ceiling, Briscoe Western Art Museum. *(Don Kirk)*

Replacing Original Windows

Pam applying a glaze to the ornamental plaster on the ceiling of the chapel at the University of the Incarnate Word. (Tommy Adams)

Windows are a big component in the restoration of architecture. Often people want to replace old windows with what is readily available on the market, which, unfortunately, has usually been aluminum or anodized aluminum windows.

Happily, in the last five years, the window manufacturers have come up with replicas of historic windows that look like the originals, though, of course, it's not the same as truly preserving the original windows.

On the projects that my husband Lloyd has had—like the Maverick Building—they restored the old original windows. We think if the windows are intact and can be restored, by all means hire John Hall, because he is very good. He's probably the best.

But, at places like at Fort Sam Houston, in the old BAMC building and Beach Pavilions, the windows weren't wooden; they were the old metal crank-out casement windows. They wanted to put in some energy efficient windows. They didn't do it for the tax credit (as the U.S. Army, it's not relevant), but they wanted to upgrade and save energy. So they put in a replica of whatever was there. I don't know if it pushed out; I don't think it cranked. But you couldn't even tell the difference.

The State Capitol Doors Project

The Jary's work on the Texas State Capitol was important because of the historical value.

> *There were 125 document vault doors. Out of that number, all but thirteen to fifteen had been sandblasted. All the paint removed, everything! All ornamentation gone. The Preservation Board asked us to come in and document the vault doors, which meant taking the existing doors that had been painted over and taking the paint off layer by layer until we exposed the ornamentation. They were all deadbolt safe and lock doors. Each one had a pastoral scene painted on it, the name of the state agency and a gold leaf band that went all around it.*

Brushing on copper powder. *(Tommy Adams)*

> *The time it takes to do a job like this just depends on how many layers of paint there are. With two of us, it took less time. First, you expose a small area so that you know what your procedures are going to be. There were eight layers to remove; it probably took a week to really clean the doors.*

Once again, Jary was performing 'paint archeology,' the removing of layers of paint, layer by layer, to determine the history of the doors.

> *Then, we put a temporary protective layer on the doors, because they didn't have the money at the time to restore any of it.*

The Conservationist as Scientist

> *There is a lot of science in what we do. You have to be very careful with the material that you use in conservation so that one chemical doesn't cross link with another and cause a problem. So, we're very careful about avoiding certain combinations of chemicals. For example, you can't use the same chemical on a plaster wall that you would use on a metal vault door.*

(Don Kirk)

This 1926 Spanish Colonial theatre was razed in 1983 with only the south facade on Houston Street preserved and restored. A bank thought their high-rise building was more important. The bank went out of business shortly thereafter.

Historic Theatres

The Jarys loved working on the Empire Theatre in San Antonio and felt it was like working in a little jewel box.

> *All the gold leaf and copper powder and all that ornament that is there, now, once again, is visible.*

Early photo of the balcony boxes in the Charline McCombs Empire Theatre showing the relief plaster restored by Jary. (UTSA Libraries Special Collections)

> *The man that originally did all of the ornaments there—I think his name was Thomas Piatti—was an Italian immigrant and he cast everything that's in the Empire Theatre. Mike Battersby, a stuccoist, cast the pieces we needed for the restoration.* [His story is in Chapter 3.]
>
> *If you have a pattern book from Chicago, you can see almost every pattern that's used in the Empire Theatre. Even more amazing, they were put in there not just for ornamentation, but this was how they did acoustical technology at that time. You had all these service levels then and that's how they created their acoustics.*
>
> *At my initial job, the Galveston Opera House, there was no echo in there originally, but they had ornamental plaster then that's gone now. When they put in the new audio system, there was a bad echo so they had to try to deal with it.*

The Alameda Theatre, built in 1949, was one of the last grand movie palaces built in America and the largest theatre ever dedicated to Spanish-language films and the performing arts. (Don Kirk)

Pamela applying glaze on the ceiling of the University of the Incarnate Word Chapel. (Tommy Adams)

In the late '20s, when they built the Majestic, they came in and painted over that ornamentation at the Empire, but they used tempera paint, so when we were doing our paint research, all we had to do is take a sea sponge and water and wipe it off to see what it used to be. But, I will say one thing—San Antonio is very special as far as historic theatres go. We have the 1914 Empire Theatre built during the golden era where everything was done in gold leaf. Then, you have the atmospheric theatres that we built in San Antonio like the Majestic Theatre, the Aztec Theatre, and others we used to have, like the Texas!

And then we have the Alameda, which is post WWII, and everything in that theatre is still intact. There are very few cities that have theatres from four periods in history that are still preserved. With the Tobin Collection at the McNay Museum, we absolutely have a serious theatre presence in San Antonio.

Family Background and Influences

Cisi was not supposed to become an artist, though, she says

I had an Aunt that was a school teacher and she was sort of an artist and I went to Incarnate Word as a boarder and took art as a youngster from Sister Maria Del Socorro.

I'm from a small town south of Alice and Kingsville. I'm from a ranching family. My father, Gus Canalas, was blind, but never let that stop him. He was County Commissioner down there and was defeated when that Box Thirteen incident happened. Dad died in 1976, very young—at sixty three—of lung cancer; he smoked non-filter cigarettes.

Cisi's mother is ninety years old as of this interview and lives in San Antonio. Her brother is in ranching.

My sister really was Dad's eyes. Because she was blind, she worked in his office with him. I didn't start the company until '85 and so my Dad never knew any of my work. Mother has, and my family has, but Dad never saw it. But he was always famous for saying 'You'll never make a living as a painter.' And every time I sign a contract I fondly think of Dad. I think, 'Here I am, Dad, making a living.'

When I first went off to college in Florida, I was supposed to be a pre-law major. Because I had an uncle that was a judge and, of course, my Dad, I was supposed to be a lawyer. My sister was supposed to be a commercial business major; my brother was supposed to be a geologist.

I just didn't do it, and I didn't tell him until I got back. And then, when I got back, I had been accepted at the Art Institute in Chicago. I thought I would really get to go there, but since he was paying my tuition, it was either go to UT Austin or not go at all. So I chose to continue with my schooling. And that's where I met my husband, so it was good.

Analyzing a paint chip in a glass petri dish. (Michael J. Smith)

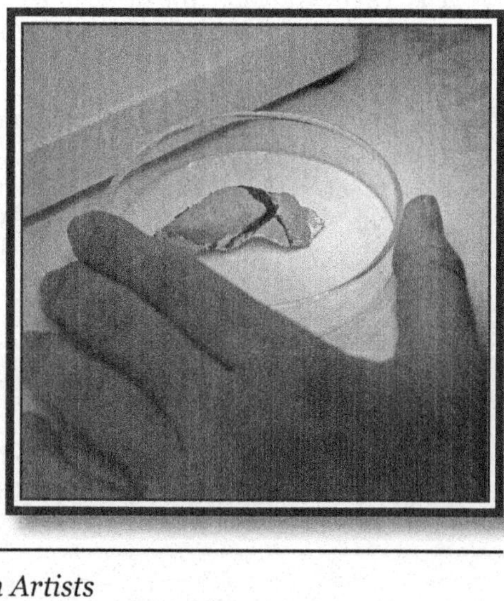

Influential Teachers and Classical Training

Luigi was my main influence and a mentor for me. He was just a traditional classical painter, and he demanded that of his students. He went to the University of

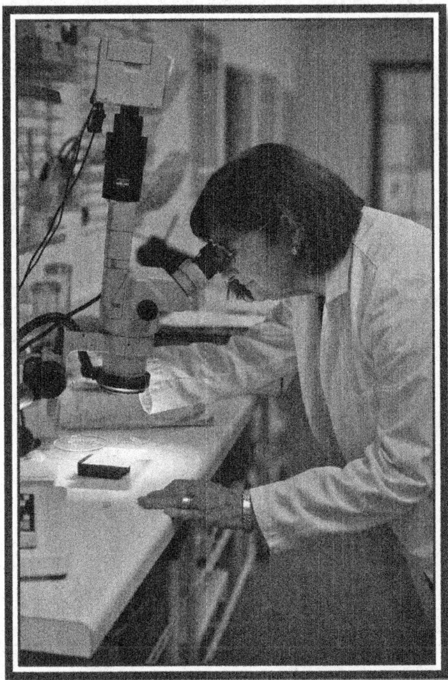

Cisi Jary analyzing a paint sample.
(Barbara Dean Hendricks)

Colorado and taught there for many years. There's a place in Philadelphia now where you can go and take classical painting, but it's one little school. I think students who don't get exposed to it miss a lot. I went up to A&M to lecture to some graduate architecture students, but they like to do everything on the computer now. They aren't required to do elevations with watercolors so a lot of the architects don't ever do an elevation other than on a computer. When I showed them a tube of paint some of them didn't know what that was! You know the classic thing is 'Haven't you drawn on a napkin yet?' My husband is my age, and in all of their classes they had to draw and build models. They still build study models, but they're not as detailed. It shows in the elevations; you know some of the architects of the past were artists. They could do beautiful elevations. Not anymore. It's so sad. It's simply easier to 'draw' on the computer. And they are nice. I'm not saying they shouldn't do it on the computer. I just think they need to do both or it's a lost art.

Hand-Eye Coordination

In Jary's field, hand-eye coordination is a critical ability.

And in some art classes they will tell you to draw with your left hand as well. In one drawing class, if you were right-handed you drew everything left-handed. The main purpose was to improve hand-eye coordination. It makes you see things a different way. Another class assignment was to draw a copy of a famous drawing, but you had to turn it upside down so your brain could not identify if you were drawing a hand or a foot or whatever and make assumptions.

(Don Kirk)

San Fernando Cathedral, San Antonio, Texas
Built in 1750, it is the oldest cathedral in the United States.

Another Mentor

Charles Umlauf was another mentor that I liked very much and learned a lot from as well. He was my sculpture professor and, you know, he had no degree. Not very many people know that. He was from Chicago and learned drawing by going to medical school exhibits. He learned parts of the anatomy that way. You could talk to him and say 'I'm having trouble doing this,' and instead of telling you how to do it, he would tell you go look at one of the masters. You interpret what they were doing, see how they solved a problem, if it was the perspective, or texture, or whatever.

Doing Research

Both of Cisi's main mentors stressed research.

I'm absolutely hooked on doing research. It's fascinating and I think it's an essential part of growing in my profession. When you see something, if you're not curious enough to want to know why they did it that way, then I guess research wouldn't be as important. But we're all curious; I think it's one of the major human characteristics. Like most animals, we all like to explore.

Professional Associations

Jary rates these associations very highly as professional resources. The American Institute for Conservation of Historic and Artistic Works (AIC) has a quarterly journal, conventions, and conferences.

There are several areas in the American Institute for Conservation. There are architecture conservators like we are; there are fine arts conservators, and painting conservators. There are works on paper and wood conservators. This probably hasn't formed yet, but in the future there probably will be another group that will be restoring our twentieth-century materials. These are not very interesting to me, you know, like ceramic tile that came out of the first McDonalds!

It's all going to be historical some day. That's what I'm saying about all this modern architecture. For me, it just doesn't have the lasting character that the old stone buildings and brick buildings had.

Other useful organizations: the Association of Preservation Technology (APT) and the American Institute of Architects (AIA).

"As far as skill is concerned, you can't just squeeze out a tube of paint and think it's going to match. And you need to learn how to handle oil paint and watercolor." —Cisi Jary

The coffered ceiling of the San Fernando Cathedral, San Antonio, Texas. Over 300 colorful 7ft by 3ft stenciled patterns had to be restored. (Don Kirk)

"It is so important for young people to know their heritage, and to know what skills we have in this country." —Cisi Jary

The Importance of Sharing

They are resources if you run into a problem you can't solve. Also, if a student needs to know something, they need the resource that they can call and say, 'You know, I'm having a problem here. What is the usual methodology in the American Institute for Conservation?' We share information like that and I think that's a good thing because you don't have to reinvent the wheel. Plus, you have to stay within certain guidelines in conservation.

Cisi knows some conservators who play their cards really close to their chest and are not willing to share a thing.

That is actually addressed in our code of ethics.

A restored water-damaged ceiling mural in the domed entrance to the Landa Library in San Antonio. Curling paint layers and missing parts of the design were restored by Jary. (Don Kirk)

Code of Ethics

The 'American Institute for Conservation of Historic and Artistic Works' (AIC) is basically the governing body of restoration artists. One example of the code includes sharing information. Also, a conservator is not supposed to put a value on historic work; in other words you're not appraisers. People will say, 'Can you put a value on my painting? Can you put a value on my sculpture, on my furniture?' We say, 'No.' Our job is to save them, not appraise them.

Restoring Statues

One of the statues we worked on was in the St. Anthony church on Lorenz Road. It was a statue of the Virgin Mary. Her whole blue cloak had silver under it, but it was all painted blue. The way that was done, they would completely gild the surface and then they would put a stencil pattern on the top that made it look like lace. Sometimes they would use silver, sometimes gold.

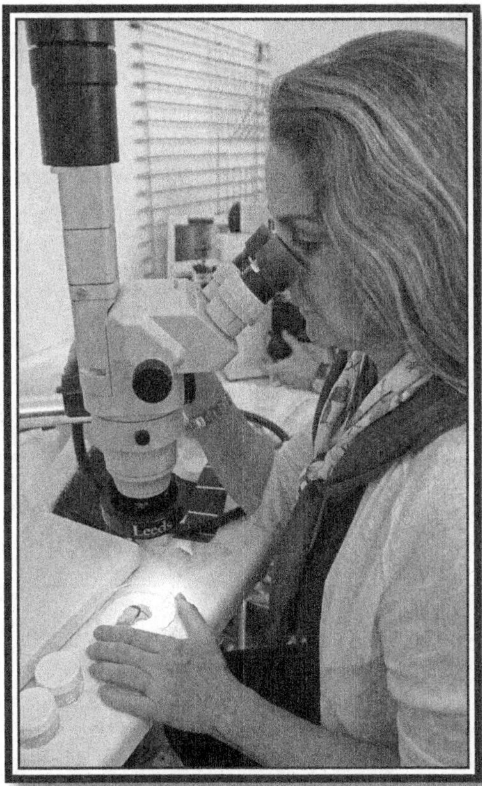

Elisa analyzing paint chips. (Michael J. Smith)

> *I'm sure you've seen statues that have white borders, well they put the gold on first as a solid band and then stencil the color on top of it. And if they sometimes used real lace, they put a base color on it and stencil the pattern on top of that.*

Every Job Has a Story

> *Everything, absolutely everything, has a story. You do research.*

She shows us a special ceiling section.

> *That's a piece of the original ceiling from San Fernando Cathedral that we took down and numbered and then reproduced with some acoustical and heat requirements. As you can see, it has a frame underneath and there is Styrofoam; that's how they released the heat factor because heat was affecting the paint up there.*

Creativity and Being True to the Past

Cisi is quick to differentiate between contributing to original works, effecting changes, or translating another's art and doing restoration work.

> *You get to use your artistic skills, but, you don't get to create. In fact, before we ever start a project we always tell the artists that are working for us there is no artistic license allowed here. The work is really not ours. It belongs to whoever the original artist was. We're only trying to preserve it.*

She is occasionally asked to create original pieces, but mainly her creativity is in service to the original artist, being true to the past.

The entry lobby ceiling of the McNay Art Museum in San Antonio, showing restoration work by Jary's Restoration Associates Limited. The task included cleaning, painting, replacing missing elements, and applying protective sealers. Other rooms were restored, included the dining room and the library ceiling, all meticulously performed by working over one's head. (Don Kirk) (Below, Courtesy Restoration Associates)

The work at the Empire Theatre involved mostly restoration of what was historically there, but had to be recreated. So, after we did color matching, that was all we did. Usually we don't sign it, because the work is really not ours; it belongs to the original artist even if we don't know who it was.

She gives use some examples:

At the Steves Homestead, we had to do a lot of painting. I think the original artist didn't sign the work, but he painted design elements that were shadows of leaves—and the same design element is at the Women's Club. So I think instead of him signing his name, he used these design elements to indicate that it was his work! The Landa Library here in San Antonio had no signature, but the Navajo Nation Council House <u>was</u> signed.

Whatever the case, Jary and her daughters are following in the tradition of the medieval artist whose work spoke for itself. While the Jarys don't sign their work, they do keep meticulous records.

We do photographic documentation that we give to the clients to put in their files. If they want to see who did the work they can look in the notebooks.

On the City Council Chambers job, we did the Crawford ceiling and the mezzanine with all that ornamental plaster. There's probably twelve or thirteen volumes of notebooks that we have for each particular coffered ceiling, in which we detail what the procedure was and what it required.

The "Ghost" in the Building

This is going to sound really kooky, but in some buildings, you can feel that the energy of the artist is still there. At the entry in the Landa Library, they used a pattern that they thumb-tacked into the plaster, and the little tack holes are still there, and so are the fingerprints of the artist. At the Moody Mansion in Galveston, you could just feel the energy of the artist that was there. At St. Mary's Catholic Church the same way, and St. Joseph was the same.

Ironically, at the Galveston Opera House, others had known of an opera house ghost, which I never saw, nor did I ever feel the energy of the painter. But in these other places, where a lot of the original work is still there, it's a very strange sensation, it's eerie!

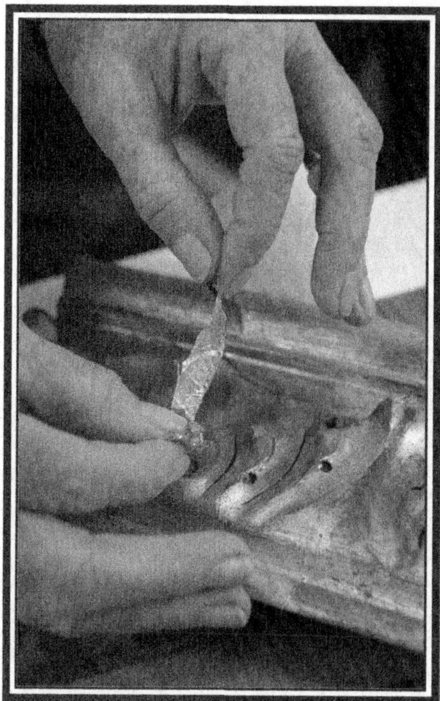

Applying gold leaf. (Michael J. Smith)

The Skills Needed to Enter the Field

Besides curiosity, Jary listed a number of qualities and educational areas she believed would be helpful to someone wanting to work in the conservation/restoration field.

Art History is important, and she recommends studying the masters.

> *Also, they do need some painting skills. They need a few classes in architecture and they must have courses in the History of Architecture.*

She says they should be able to answer the question 'How do they build these buildings?'

> *They need to know the different styles of architecture so they can identify Gothic to Renaissance to Vernacular styles. We have outstanding examples of all of these styles right here in San Antonio. And know the types of windows and types of arches and their dates, because those dates can be beneficial to your research. Even the way brick is laid is important, but, you don't learn that in an art school.*

There is a lot of on-the-job training necessary to be successful.

> *As far as skill is concerned, you can't just squeeze out a tube of paint and think it's going to match. You need to learn how to handle oil paint and watercolor. Acrylics would never be used in a conservation treatment; it's a form of plastic.*

"This probably hasn't formed yet, but in the future there probably will be another group that will be restoring twentieth-century materials. These are not very interesting to me. You know, like tile that comes out of the first McDonalds." —Cisi Jary

It's not going to age and it's never going to have a patina like historic paint would.

Then a little chemistry is helpful, to which I was fortunate enough to have been exposed. I wasn't very good at it; that's why I didn't stay with it, but the workshops I've had have been very useful.

Where to Study

For those considering this field, Cisi thinks they should start in art school.

In San Antonio, Trinity and Incarnate Word all have good art departments. UTSA produces especially skilled artists and art teachers. Texas A&M has both Conservation Architecture and Preservation Architecture. UT Austin has a department, but it's not as big as A&M. From there, if someone really wanted to go into the field, they should go to the Campbell Institute or they should go to New York City University, which has a Preservation/Conservation Department.

A ghost in St. Mary's Catholic Church? (Don Kirk)

Jary's Legacy

I guess when I can't work on a scaffold anymore, maybe I'll retire and write my book.

She says that even now when she remembers something, she "puts a memo" in her computer.

As disjointed as that may seem, it's the best way I know to remind myself of something I just came across.

The Competition and Future of the Company

There are some firms and a company in Dallas; there's two that I know of in Houston, but I think that's about it.

She believes the business will continue under her daughters.

I hope so and I think it will, I really do; I think they love the work so much now that they've done so much of it. And every project is different. That makes it exciting and I don't think they will be willing to give that up.

Lobby ceiling of the 1929 Art Deco San Antonio Express-News Building. Restoration work was done by both Jary and Battersby. (Don Kirk)

Survival of the Field

Cisi believes the field will survive *because of the public interest in, and respect for, historic buildings. It's created a tourism destination in the United States that we didn't have before – unlike Europe. It's very beautiful work and the craftsmen who created it in this country were mostly immigrants. They learned their skills in Europe and came over here as immigrants. That in itself is probably another whole period you could research. Because of that, I think a lot of young people may go into the field, but perhaps not stay in Texas because there's more restoration work on the East Coast. The Boston Opera House, for instance, and there are all these different 'atmosphere' theatres throughout the eastern U.S.*

Importance of Protecting Our Heritage for Future Generations

It's important, not just for San Antonio, but everywhere in the Southwest. San Antonio really is advanced. If you go west—Wyoming, Idaho, Utah, all of those western states—they've only just begun as far as their preservation architecture is concerned.

Jary's restoration of Scoates Hall ceiling, Texas A&M University. (Rick O'Neill)

They've just begun and they have beautiful state capitols, they have beautiful county courthouses. Salt Lake City has a lot of historic buildings in the downtown area, but they're just a little behind.

In the Midwest—Kansas, Oklahoma, Texas—we were behind the East Coast. They actually started doing restoration before we did. And then it sort of moved east. San Antonio has been fortunate because we've always done it. The Conservation Society has always been involved in preserving historic buildings. And usually, when they attempt to preserve these buildings, they don't usually disturb the walls too much. The decorative painting is still there—maybe under eight or ten layers of paint, but it's still there!

I hope the public creates enough interest so there will always be a need for architecture conservators and painting conservators. Painting conservators—as opposed to repairing single paintings—there will always be a need for that. But wall art, and murals, and ornamental plaster, I hope they continue to restore them.

It is so important for young people to know their heritage and to know what skills we have in this country. I mean, how did we get to where we are today? In Europe, people save everything; they have such respect for artisans with a skill. We do here, too, but we're a wrap-up-and-throw-away society. A lot of times we go on a job and they want it restored, but they want it done in thirty days. That presents a problem. Usually we are given more time when we tell them you can't make paint do what you want it not to do.

If you look around at modern structures now—and this is a standing joke I have with my husband—I ask, 'Which one of these glass-walled buildings are they going to want to save in twenty years?'

That about says it all. Cisi and her daughters can be proud of what they have helped to preserve: historic hotels, theatres, state capitol buildings, courthouses, Spanish missions, places of worship—a long list of restoration projects that will, just possibly, now last into the next century for our great-grandkids to enjoy.

3. Thomas Michael Battersby

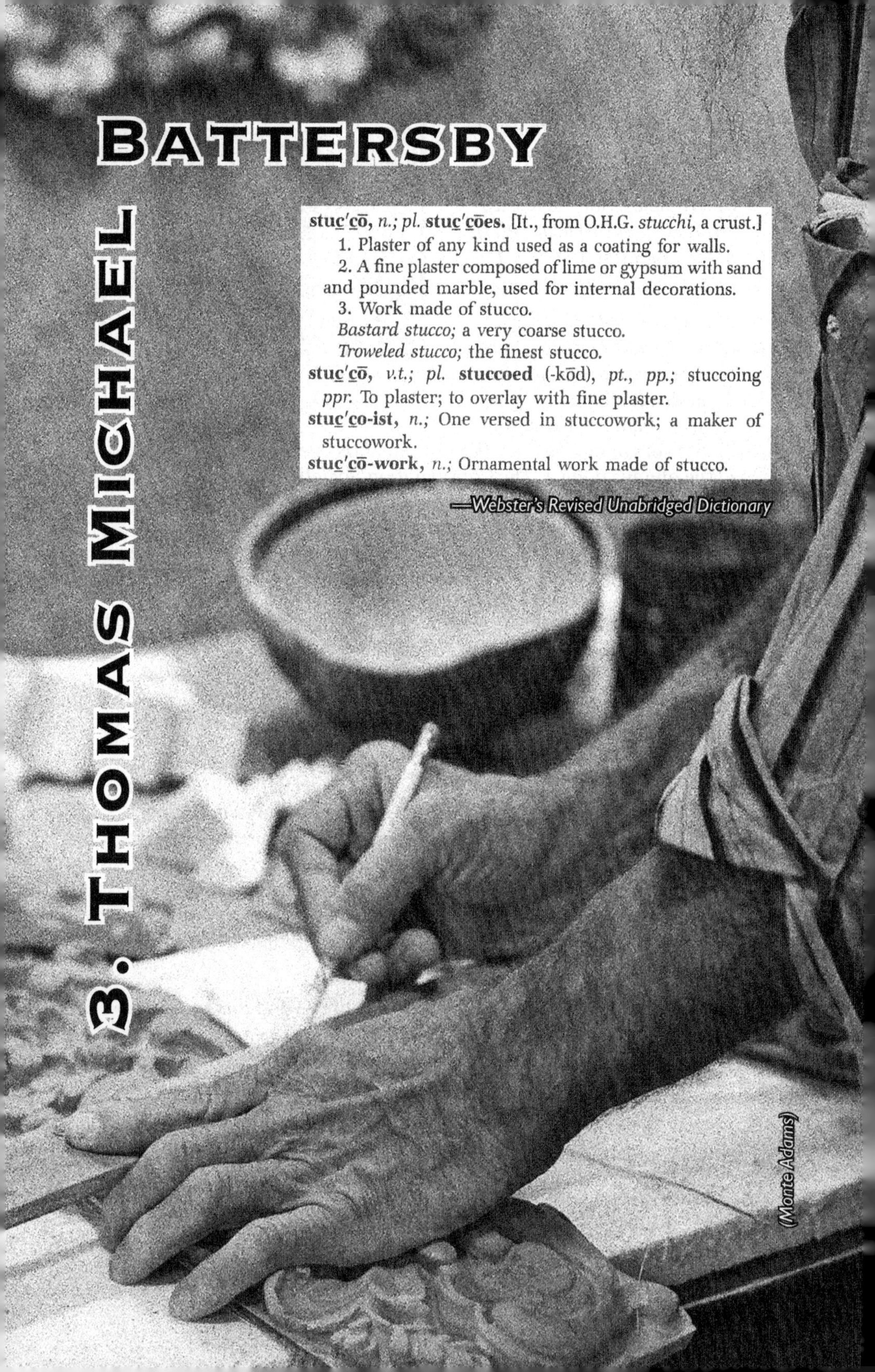

stuc′cō, *n.; pl.* **stuc′cōes.** [It., from O.H.G. *stucchi,* a crust.]

1. Plaster of any kind used as a coating for walls.
2. A fine plaster composed of lime or gypsum with sand and pounded marble, used for internal decorations.
3. Work made of stucco.

Bastard stucco; a very coarse stucco.

Troweled stucco; the finest stucco.

stuc′cō, *v.t.; pl.* **stuccoed** (-kōd), *pt., pp.;* stuccoing *ppr.* To plaster; to overlay with fine plaster.

stuc′co-ist, *n.;* One versed in stuccowork; a maker of stuccowork.

stuc′cō-work, *n.;* Ornamental work made of stucco.

—*Webster's Revised Unabridged Dictionary*

(Monte Adams)

INTRODUCTION

BATTERSBY calls himself a "stuccoist" and is the third generation of artisans working in plaster, molds, and casting in San Antonio and South Texas.

Battersby frequently finds himself preserving the creations of his father and even his grandfather, the first plasterer in the family. In the 1980s, Battersby and his father worked together on a major restoration of the Majestic Theatre, which, by the way, his grandfather had helped build.

Some of the important public buildings Battersby has worked on include the missions, the Aztec Theatre, and the historic Menger Hotel. His work requires many skills: welding, carpentry, tool making, and mold making among them. The job requires all kinds of troubleshooting and problem solving. And while restoration is most of his business, he loves to work on original designs.

Battersby's favorite tools are his hands, but next in line are his father's and grandfather's tools. One of his favorites is his grandfather's old square that he cut down. He makes many of his own tools, whatever is needed to accomplish the task.

Like the other trades (woodworking, masonry, and blacksmithing) Battersby says his craft is as old as time itself: "Working with mud has been around a long, long while."

(Monte Adams)

Thomas Michael Battersby.

Thomas Michael Battersby

Stuccoist & Lead Beater

Thomas Michael Battersby, who prefers to be called simply "Battersby," is the son of Thomas Richard Battersby (a plasterer). Battersby was born in 1949 in San Antonio, where he has spent most of his life. The Battersby family emigrated from Great Britain in the latter part of the 1800s. The first Battersby in America was a dairy farmer. His son, James, was a blacksmith. James' son, Charles Richard Battersby (Michael Battersby's grandfather) was the first plasterer in the Battersby line. Battersby says:

> *I don't have any children, so I am going to take the plastering trade as far as I can while I'm still kicking.*

Battersby's grandmother often told him about James, the blacksmith. When James needed money, he'd just make some tools and take them by the side of the road and sell them.

> *I always thought, 'What a life!' That'd be great if you could exist like that today.*

The infamous dog "Blackie." (Battersby)

Battersby's dog *Blackie* is an important part of his life:

> *She is going to tell you that she's not mine, that I'm hers. She found me about ten years ago. I was out on a job on the northwest side of town. She was pretty young then. She came running through, running wild, and my previous dog, that I had for a long time, had just died six or eight*

months previous. The old guy that I was doing that job for saw Blackie come running by and I said, 'I ought to take her home' and he said, 'Yeah, you ought to, she's been running around here loose and lost for two or three weeks.' So I called her over and she came up to me and she got in the truck. She came with me and just took over when she got here. Blackie, good girl.

Battersby has numerous tattoos on both arms, some dating back to the 1960s. Fifty years' worth, he says.

They've become a lot more popular now. It seems like everybody has them. My sister-in-law has one. Her twin sister has one. My nephew has one. Lots of people have them, but I don't recommend it, I tell you. You know who is really fascinated by them? When I go see the doctor, he's always calling the nurse in and saying 'Come in and look at this.' Maybe that's why I got them.

Battersby's dog Blackie and his tattoos are captivating, but Battersby's incredible craftsmanship makes the artist even more fascinating.

I like to meet good people, and, like Blackie there, I can tell good people when I'm around them for a minute or two.

Plaster artisans working the Neisner Building in 1947. Battersby's father Tom is on the right. (Battersby)

We can tell good people too, and Battersby, descendant of a long line of artisans, is good at many things—not the least of which is creating plaster-work pieces that are genuine works of art. Battersby has followed in his father's footsteps. He calls himself a *stuccoist*:

> *When you tell somebody you are a plasterer, they think sheetrock. There is nothing wrong with sheetrock, in its place, but I'm a stuccoist, which really is a wider range of skills. I am a mold maker, I do casting work, and installation of the work as well.*
>
> *The work that I do requires a lot of different skills: welding, carpentry, all kinds of things to make molds, but with my grandfather and my father in the trade, I have been around it all my life. I really started learning it when I was quite young. A big part of what I know and do is from my dad and my grandfather, but Dad also taught me to move on and expand into different areas of the trade. I do a lot more mold making than he did.*

Trimming plaster castings for the Majestic Theatre in San Antonio. (Monte Adams)

In 1994, Battersby restored the coffered ceiling of the San Antonio City Council Chambers in the former Frost Bank building on Main Plaza. (Cisi Jary and Pamela Rosser re-gilded and repainted the ceiling.) Battersby also created a new larger, plaster casting of the city seal based on Johannes Scholze's design. It now hangs on the wall behind the Mayor's chair. (Courtesy Battersby)

His Work on San Antonio Buildings

Battersby worked on a number of important San Antonio buildings, including the missions. His comments on the importance of saving San Antonio's historical heritage are illuminating:

> *Thanks to some folks that care about those kinds of things. The Conservation Society, of course, is a good way of saving this stuff. Joci Straus is a big top player in saving some of these beautiful buildings. The building I am working on right now is the Majestic Theatre, for example, which almost fell to the wrecking ball ten or twelve years ago, and I guess it was the Conservation Society along with Joci Straus and a bunch of other people who saved it. The list is long of the people who gave blood, sweat, and tears to save the Empire Theatre, the Aztec Theatre—which is still waiting for work—and the Majestic.*

Battersby spent about a year on the Empire Theatre, *"rebuilding, restoring a lot of it,"* and he told us about the faces in his studio that came from the Empire Theatre.

> *I just call him the ghost. They are around the box seats. The band that goes around the box seats is where those three faces are.*

He also did some restoration work on the old Frost Bank building, which now houses the city council chambers. He reveals the "inside story" on one facet of its restoration. Pointing to a plaster flower in his studio, he says:

> *In fact, that flower right up there is from that ceiling . . . It's a coffered ceiling. I counted them one time, there's like a hundred and sixty-three or a hundred and sixty-nine . . . There is a bunch of coffers up there. They are way up . . . I think it's about thirty-five feet up to that ceiling in the old Frost Bank building.*

A three-foot-diameter "flower" casting hanging on Battersby's workshop wall. One of over 160 used in the coffered ceiling of the San Antonio City Council Chambers. (Don Kirk)

The Majestic Theatre grand opening in 1989 after extensive restoration. (Don Kirk)

Majestic Theatre

Battersby and his father did a major restoration on the Majestic in the late 1980s. He speaks fondly of that time:

> *We were really all over that theatre from top to*

bottom, from stem to stern. When they did that restoration in '89, and, as you are aware, they did everything—brand new carpeting, brand new seats, new paint on everything. My dad and I, every little bit of plaster, they told us, 'just fix every little nick.' We spent right at a year in there, close to a full year, and we just went from top to bottom. From the ground up to the top at seventy feet, they put up massive amounts of scaffolding so that every trade on that project could get to every square inch of the theatre. It made it convenient for necessary fixture changes, repairs, and restorations while everybody was there and it was really well organized. They hired the best people to do the architecture, the planning of it. The big theatre people—people who knew what they were doing—got in there and ran it. My dad and I had the opportunity to work on the plaster restoration. Of course, they had an excellent carpentry crew. Everybody really put their heart into it. It was a really great job to be on.

Michael's father, Tom Battersby.

This photo was taken by John Dyer in 1990, about two years before Tom's death in 1992. At that time they were working on the Linda Pace residence designed by the late, great architect Isaac Maxwell. (Photos Courtesy Battersby)

That's Battersby in a Bosun's chair, hanging precariously over the proscenium arch in August of 2006 during a restoration of the theatre his father called the "Magic Stick."

San Antonio's Majestic Theatre, in the "Atmospherics" style, was designed by John Eberson and built by Karl Hoblitzelle in 1928. The design attempted to create the illusion of an exterior space in an interior venue. It was saved from the wrecking ball in the 1980s.
(Photos Courtesy Battersby)

Battersby is proudly holding the San Antonio Conservation Society's Lynn Ford Craftsman Award presented to him and his father in 1990 for "exceptional skill in restoring the ornamental plaster in the Majestic Theatre, San Antonio, Texas." The three objects at the top of the trophy symbolize the three mediums that the late Lynn Ford worked in: plaster, wood, and metal. (Courtesy Battersby)

Battersby reveals the secrets of the clouds that move on the ceiling of the Majestic:

They call it a brenograph [projector] *that shoots the light up there. It's got patterns cut in a circular disc that rotates and makes the lights look like clouds moving across the ceiling. And, of course, the stars are up there permanently in the ceiling.*

His Father and Special San Antonio Memories

My dad always called it the 'Magic Stick,' rather than the 'Majestic,' and around the corner was the 'Impure' [Empire]. *My dad passed away in '92, so he didn't have the opportunity to be along with me on the restoration of the Empire Theatre, which was built in 1913 or 1914. The Empire showed the 'B-rated' movies . . . It's a lot smaller, of course, it's only nine hundred seats, I believe. It's much more of an intimate setting at the Empire.*

The Majestic had twenty-eight hundred seats, I think. It was built in 1928, the year my dad was born, and my grandfather was working at the Majestic when they built it. The Majestic had not only the 'magic' of the theatre, but they also had the best movies and entertainment, what have you. Yeah, the Majestic everybody remembers—everybody loved the Majestic Theatre. My dad always said that the highlight of his career was to get to do the restoration of the Majestic. That was really a great job!

A two-faced Mayan mask above the River Walk entrance to the Aztec, 2005. (The Di Filippos did the Terrazzo work on the pavement.) (Battersby top two photos, Don Kirk bottom two)

AZTEC THEATRE RESTORATION

Condition of a fire-exit surround of the Aztec before restoration. When the theatre was divided into three theatres in the 1970s, the original surrounds were destroyed. (Battersby)

The drawing Battersby worked from to carve the new surround. Drawn by Ricky Armendariz, 2005. (Don Kirk)

Battersby standing in his workshop next to his carved plaster surround before moving it to the Aztec in 2005. (Courtesy Battersby)

THE NEW FIRE EXIT AT THE AZTEC THEATRE, 2005

(Battersby Photos)

The Menger Hotel

Battersby restored the ceiling in the lobby of the Menger Hotel in November and December of 1991.

> *That was a great little job. It's nextdoor to the Alamo and I always like to walk in there and have a look at that lobby ceiling. That's another one where they did a real nice restoration. Lewis Fisher was the architect on that one. He's a buddy of Milton Babbitt, who did the Empire and the Majestic and a lot of others, of course. Most of the architects that I have had the opportunity to work with are easy to work with and are good guys and I have to say Milton's at the top of the list. He really is. He's so easy to work with, he's so talented; he's given me a lot of opportunities to show what I can do.*

Copper profile tool built by Battersby for use in creating the plaster cornice in the original Menger Hotel lobby. Below, a detail of the finished cornice, 1991. (Don Kirk Photos)

Work Philosophy, Tools, Secrets of the Trade

Quality is of primary importance to Battersby:

> *When I was young my dad told me, 'If you're going to do it, you are*

going to do it in such a way that when you get through, you write your name on it so everybody can see who did it. Otherwise don't even bother to do it. Don't do something that you're not going to want to write your name on.'

That's pretty much why I don't get in a hurry. Not that I have tried to slow anything down, but if things aren't going right, I just step back for a minute and take a look, because when I get done, I want to be able to write my name on it and let everybody know that 'Battersby' built this.'

Research

Battersby learned from his father and grandfather and otherwise is self-taught, but he does do a lot of research for a project.

There's not a whole lot of written information on mold making, but it's getting to be more available due to the Internet. I use the process a lot. A lot of my old tools that I use occasionally have to be replaced and sometimes they are real hard to find, but the Internet will allow me to find some of these makers—these Italian tool makers, German tool makers, etc., that are not so easy to find.

The Tools

Battersby's favorite tool is his hands, but next in line are old tools handed down by his father and grandfather. A particular favorite is one of his grandfather's old squares he cut down for a special job. He also uses his great-grandfather's vise and tongs.

James was a blacksmith and that's his 'post-vise' or 'leg-vise,' whatever people call it. There's no telling how old this vise is. I've got his anvil outside, a 160-pound anvil!

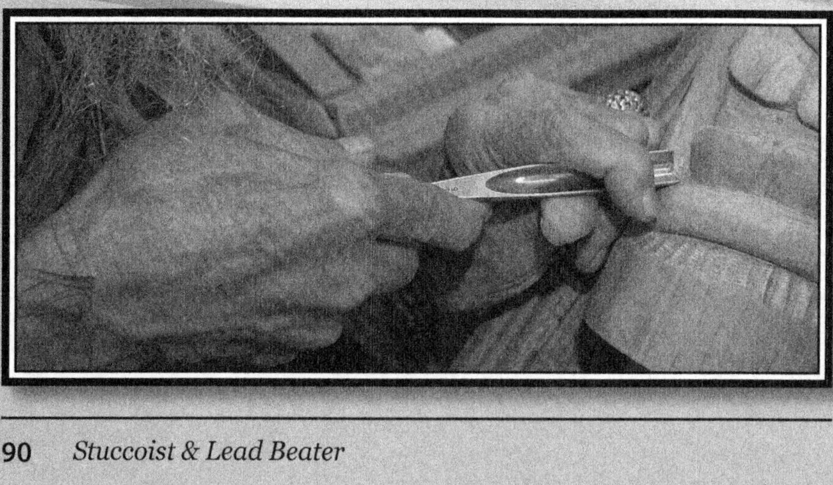

(Don Kirk)

Michael repaired the original plaster work throughout Hotel Havana and built up new veneer plaster to the lobby ceilings and corridors in 1997. (Don Kirk)

He was using the vise the day of the interview to bend some channel iron. He uses it all the time. The tools remind him of stories about his ancestors.

> *My grandmother used to always tell me about James, the blacksmith. When he needed money, he'd just make some tools and take them by the side of the road and sell them. And I always thought, 'What a life!' His blacksmith shop was down by the Medina River out south of town. They talk to me, the tools. They have already learned from my grandpa and others. They already know what to do; all you have to do is pick them up and they just seem to do the work for you.*

He says that, in his trade, many of the tools he uses he has to make himself.

> *Either you don't have it handy or can't get it at all, so you just go ahead and make it.*

Some of his tools have been handed down through generations. Tyzack, of Sheffield, England, made one particular prized tool. He says tools like these aren't made anymore.

> *They hammered out these shapes from a piece of round. They hammered it out the way they wanted it.*

And, of course, tools evolved. That's another thing his father told him:

> *Don't be scared to use these new tools. Just 'cause we are nailing it together today, don't be afraid to get that new screw gun and screw it together in the future. Take advantage of what is there, use what you can.*

Battersby has one-hundred-year-old tools <u>and</u> brand new ones—whatever is needed for the job.

A WALL OF DRAFTING TOOLS.

Above: San Antonio's city seal, in cast plaster by Battersby. *(A much smaller version was carved in wood by Johannes Scholze in 1925.)*

BATTERSBY'S PLASTER CASTING WORKSHOP, SAN ANTONIO, TEXAS

A number of different skills are required, and thus a lot of tools.

(Don Kirk Photos)

CASTING OF A LARGE BRACKET USED IN THE EMPIRE THEATRE.

BELOW: "THE FACES," HYDROCAL CASTINGS USED ON THE EMPIRE THEATRE.

(Don Kirk Photos)

A STUCCOIST AND HIS CARVING TOOLS.

Steps to the Stuccoist Process

The first step is to make a clay or plaster model. It is not always going to be a clay model. Then you make a rubber mold of the model. Third, plaster is poured into the mold.

You get it out when the plaster is set, in twenty-eight minutes.

Pouring plaster into a polyurethane mold. (Monte Adams)

Then, it is decorated:

They decorate in many different ways, might get painted, might get gold leafed.

Installation is usually part of Battersby's operation.

They are affixed in a number of different ways—whether to adhere it with plaster, or screwed or nailed; there are a lot of ways.

Once it is dry, the plaster can be drilled, sawed or cut, similar to wood, but plaster will occasionally break. It's a brittle material and I have had some guys working for me that could break a bowling ball with a rubber mallet. It's breakable. Can you hear how it rings? Like anything else, it can be broken. On the other hand, it is very durable.

It's especially important to be careful if the plaster piece has a design with elements that protrude.

Some of the points that stick out are subject to being broken off.

The Materials: Plaster & Gypsum

Battersby notes that plaster is a special material.

It is taken out of the ground and then it is ground up and cooked and it becomes a powder. They put it in bags. I bring it over here in a bag and then I add water to it, and it turns right back into the same rock it was to begin with. Gypsum is the only rock in the world that you can do that with.

Battersby has a favorite casting plaster. From the time it is mixed and poured in the mold, just twenty-eight minutes, and it's ready to go. Most of his plaster is from Oklahoma, from the United States Gypsum Company. The term "plaster of Paris" comes from some of the earliest quarries just outside of Paris, France.

Battersby uses sisal to reinforce his plaster castings. (Monte Adams)

Plaster of Paris kind of became a generic name for the material that we work with, but there are many variations of it and in this country it's found from coast to coast, north and south.

Working sisal into the wet plaster. (Monte Adams)

The Use of Sisal

Sisal is another important material for stuccoists.

Sisal is one of the reinforcements we use in plaster. Burlap is another one we use.

Sisal is a strong, hemp-based product. It is shredded for use in plastering.

I use it when I am doing my casting—in that twenty-eight minute window—from the time it becomes fluid after I add water to just the powder, while it's in the mold and it's wet. Add the water, pour it in the mold, and then I'll put some reinforcement in there, as needed.

I will either cut the sisal up into little pieces that I sprinkle in, or—if I have a larger piece—I'll lay the sisal on top and then push it down

Before:

Battersby's restoration of massive Corinthian capitals at the Brown residence in 2009. Top: Moving a restored capital into place. Center: shows their seriously deteriorated condition. Below: The capitals have been returned to their honored places. (Battersby)

After: *One of the fully restored Corinthian capitals.* (Don Kirk)

in where I want it. Maybe I'll lay another roll over here or there and then lay some pieces across the center.

Then, for lightness or for weight consideration, I'll dig some plaster out of the mold. It doesn't have to be solid, just so it doesn't get too heavy. That also helps it to dry out faster.

State of the Craft

The craft is as old as time itself. Plumbers will argue that they have the oldest trade in the world, but working with mud has been around a long, long while.

Apprentices and Longevity of the Craft

Battersby says he's never really had an apprentice, but wishes he

did have somebody that was interested to pass along some of these skills

Old photo of the original "Sweet Grapes" surround. (Courtesy Battersby)

that are really specialized to this type of restoration. There is a lesser demand for it than blacksmithing. The trade is not strictly speaking a dying trade, but it's less popular.

He says there are more stuccoists in Chicago, Philadelphia, Washington, and the Northeast, and in Europe, especially Great Britain.

There are a lot more of these older type of buildings, especially when you get around the capitol. Europe has a lot of real old buildings and in the last few years interest has really gotten a lot stronger for preservation and restoration. I don't think there is enough of the work going on around here to generate interest in the younger fellows that should be learning the trade.

Advice to Future Stuccoists

Pay attention, do what you're told, learn all you can, don't ever think you know it all, just keep on learning—learning everything you can—from what you see, what you hear. Everything you can pick up from everybody, pick up because you need all kinds of skills to succeed in this trade.

"They talk to me, the tools. They have already learned from my grandpa and others. They already know what to do; all you have to do is pick them up and they just seem to do the work for you." —Thomas Michael Battersby

Skills Needed

The required skills are diverse: tool making, plastering, mold making.

It requires all kinds of troubleshooting and problem solving. You really need to be able to go from the design phase all the way through the execution of the model, which would be modeling your design in clay. Then your clay model is done and you have to know mold making. You have to know your casting and the installation, and it requires different skills. You might have to do some welding. There is mathematics

(Battersby)

A 1929 photo at opening of

THE "SWEET GRAPES" SURROUND

Created to showcase the five-foot bronze statue designed by Harriet W. Frishmuth in 1928.

THE MAJESTIC THEATRE
RESTORATION, SAN ANTONIO, TEXAS

In 2003, Battersby recreated a sixteen-foot-tall ornamental plaster surround that contained the five-foot bronze "Sweet Grapes" statue at the Majestic Theatre. (Monte Adams)

Column detail of the "Sweet Grapes" surround at the Majestic Theatre. Pamela Rosser and Pamela's cousin Sue did the elaborate painting. (Monte Adams

Battersby test fitting a plaster casting at the Majestic. (Monte Adams)

The Majestic Theatre "Sweet Grapes" Restoration

Photo taken during restoration of "Sweet Grapes." From the left: Cisi Jary, Jeanne Albrecht, Battersby, Pam Rosser, & Michael Castillo. (Courtesy Battersby)

Polychromed and gold leafed by the Jarys. See color photo on back cover.
(Courtesy Battersby)

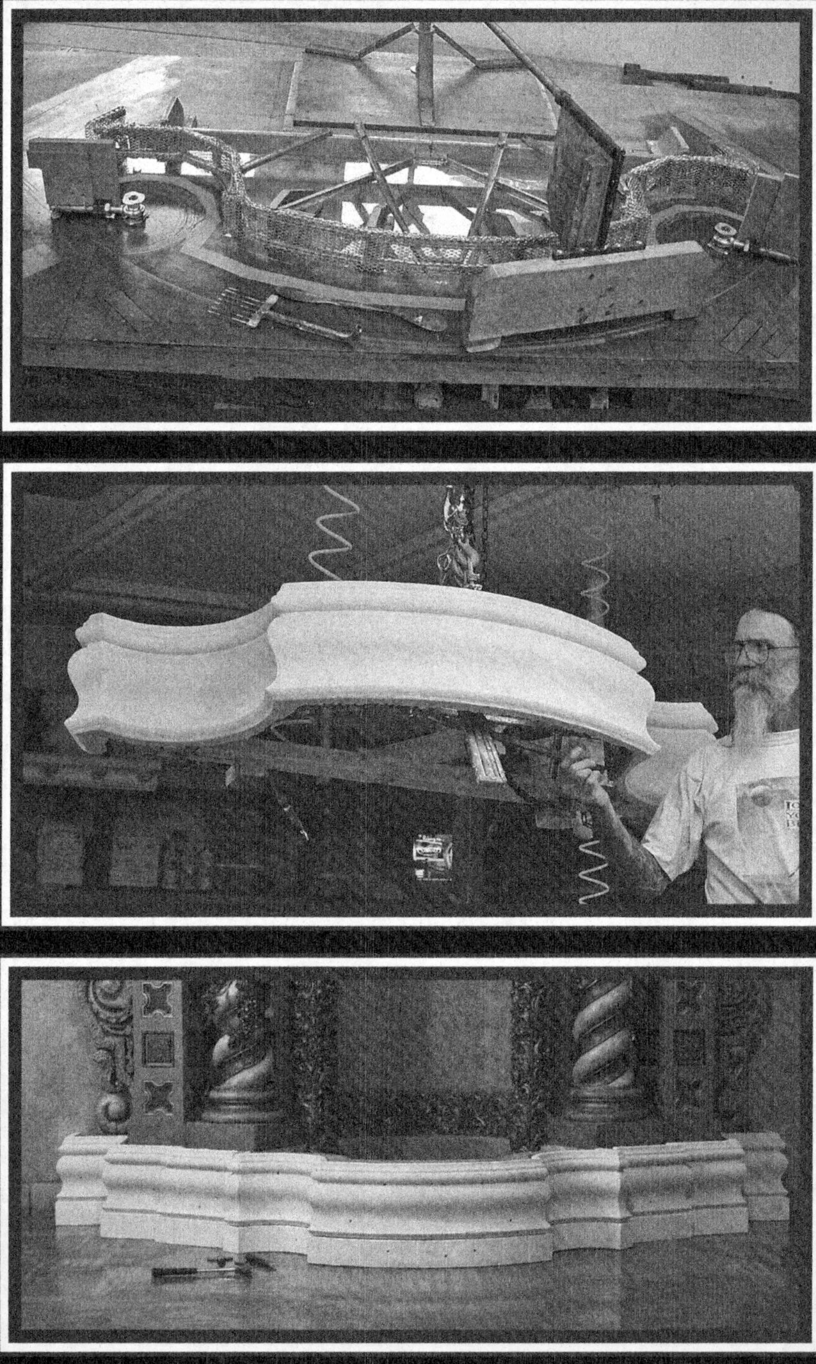

Steps in the construction of the "Sweet Grapes" base. (Courtesy Battersby)

Detail of the "Sweet Grapes" statue surround in the Majestic Theatre lobby after Battersby fabricated it from scratch. Above, showing the unbelievable sculpted detail before Cisi Jary & daughters started their painting magic. At left, the steps required to fabricate the base: building an armature of galvanized steel and metal lath, the plaster finish over the armature, and third, the base installed on site. (Battersby)

The original copper eagle, with its six-foot wing span, is again perched on the canopy of the Empire Theatre after its discovery in the ladies' restroom during restoration.

Charline McCombs

EMPIRE THEATRE

BUILT IN 1913, RESTORED IN 1998, ADDED TO THE NATIONAL REGISTER OF HISTORIC PLACES IN 1999.

A terracotta eagle medallion above, and a grotesque embellishment cast in Portland cement.
(Don Kirk Photos)

involved; geometry is really important. I go back to stuff I learned in the first grade. I use math a lot. The thing I really like about it is, it never changes. A rectangle is always a rectangle. The math doesn't change on you.

Marketing also has become very important for the business, but Battersby admits:

It's one of my worst skills.

Battersby and His Work

Battersby once recreated a ceiling for a residence in Austin that had fallen into about a hundred pieces.

I had to put it back together and make a mold out of it.

Battersby also assisted on a King William restoration with Tommy Adams and produced a piece used in restoration of the Post Chapel at Fort Sam Houston.

Original Designs/Economy

Battersby loves to do original designs; inspiration comes from everywhere. One example was inspired by an Hawaiian woodcarving, but, basically, he says:

Inspiration usually comes from the economy.

Battersby filling a mold with Hydrocal to create an "artificial stone" casting. (Monte Adams)
Below: Battersby in 2014. (Don Kirk)

One of five fireplace mantels crafted by Battersby for the Linda Pace residence, 1990. (Courtesy Battersby)

I wish I could take the time to do the things that I want to do, but it is so hard to make a living that I don't really have the time to do things I just like to do for myself. I do get to do original pieces occasionally. The inspiration usually has to come from something that is going to be paying the bills for me these days.

Favorite Projects

The last project Battersby did for Las Casas is his favorite—the "Sweet Grapes Alcove" in the lobby at the Majestic Theatre. Created in the Corinthian style in the late 1920s, the alcove was torn out in the '50s.

> *I was lucky enough to be asked to reproduce that alcove. When you walk in the front door of the Majestic Theatre, there it is. I put the first part of it there in August of 2002, and then I finished the bottom in January of 2003.*

Battersby Loves His Work

> *I wandered off and followed some other trades here and there through the years, but I came back to stucco and plaster because it's so much fun. I love my job. I work at it seven days a week. I am satisfied with where I am. The first half of my life is over now, and I am looking forward to the second half. Hopefully I will be doing the same kinds of things I am doing right now. That's one of my favorite things about my trade, you know, it's always different. I get to use different skills—and even learn new ones—to get the job done. So I love the trade. There is so much to it that it's always interesting.*

Battersby inspecting one of his "beaten lead" art objects. (Don Kirk)

And so Battersby turns to carving on his wooden Tikis and hammering out his beaten lead art objects, happy to know that someone will find pleasure in them just as he has found pleasure in making them.

Wooden Tikis carved and finished by Battersby. (Courtesy Battersby)

(Don Kirk)

Battersby has also created new works of art, and restored others, using a hammered lead process that he calls "Beaten Lead." It begins with meticulously carving a 3-D form out of wood. A lead sheet is laid over the form and the pattern hammered into the sheet of lead. Sheets are soldered together to form larger works of art.

Above: A laundry room box. Below: Computer drawing for the laundry room box.

(Photos Courtesy Battersby)

A fireplace mantlel using multiple lead panels.

Lead pencil boxes.

Below: The way a pair of Lynn Ford planter boxes looked before they were restored by Battersby.

And after restoration:

Below: A revolver's cylinder was the inspiration.

Beaten lead ashtray at right.

(Photos Courtesy Battersby)

4. CARLOS CORTÉS

INTRODUCTION

Following in a hundred-year family tradition, CARLOS CORTÉS, owner of *Tallér/Studio Carlos Cortés*, creates one-of-a-kind outdoor sculptures and functional art that can be found throughout México and the United States.

This European-based art, also known as *Obras de trabajo rústico* (rustic concrete work), was first introduced to San Antonio in 1924 by Dionísio Rodriguez, Cortés' great-uncle.

Cortés family work can be found at the Witte Museum, on the River Walk, at a former trolley stop (now a bus stop) in Alamo Heights, and many other public places in San Antonio, as well as at numerous residences. Cortés, like many artisans, grew up in the trade, learning the difficult artistic techniques for texture and color *"almost by osmosis,"* he says.

Like many others who pass their craft down through the generations, Carlos contributed to legacy family projects. In 2001, he restored the Brackenridge Park trellis bridge, which had been built in the mid-1920s by his great-uncle and his father, and then it was restored again in the 1960s by his father.

With steel armatures and concrete, humble tools like forks, palette knives, window screens, whisk brooms, and special mineral salts, Cortés creates unique shapes, textures, and colors that don't so much fool the eye as seduce it. "People think this is a lost art, but it isn't lost; it's just forgotten."

Carlos Cortés.

Carlos Cortés

Faux Bois Concrete Sculptor

On two different but equally beautiful days, one in early summer, one in early fall, we interviewed one of San Antonio's artistic treasures, the latest Cortés family member practicing the unique art of naturalistic concrete sculpture. Carlos Cortés is an undisputed master of the art and the inheritor of a proud family tradition. We sat down with him at his outdoor studio in downtown San Antonio.

I call it 'Studio Carlos Cortés.' I've been located here about twelve years. In the beginning, there was no particular name to the business. We were just known as the people who did this sort of work. I sculpt with concrete and people call it 'faux bois,' meaning 'false wood' in French, or in Spanish it's 'obras de trabajo rústico' meaning 'rustic concrete work.' Or, you can just call it 'trabajo rústico.'

A FAMILY TRADITION

History of a Family Craft

'Faux bois' sculpting is 'European' in nature, but it originated within my family in the México City area. My great-uncle (on my mother's side), Dionísio Rodriguez, knew the technique. He was born in 1891, in Toluca, México. He worked with a group of men that were doing projects in México City for civil engineers—in the parks of México City. He and my father were about twelve years apart. My father was born in 1903 in Monterrey, México. My father was working in cement and concrete and doing imitation stonework, and they had been doing this for a long time because my grandfather (his father) had a shop in northern México. My grandfather, along with my great-grandfather, did imitation natural stone faux finishes and imitations of marble and granite for some of the buildings in México.

They did ornamental cast concrete work that was architecturally precise. It's linear as opposed to organic looking. And they were

CARLOS' OUTDOOR STUDIO IN DOWNTOWN SAN ANTONIO.
'FAUX BOIS' SCULPTED PIECES AVAILABLE FOR SALE. (Don Kirk)

model makers and mold makers—creating steps, staircases, balconies, and the like. Though I have some pictures, sadly there are no examples of my grandfather's work in San Antonio.

So, my dad had a lot of experience in this sort of work when he met Dionísio around 1921. Dionísio was coming through northern México and Monterrey—headed eventually to the U.S.—and was looking for work. They worked together on a particular project and I believe my dad first began to be influenced by my great-uncle's work at that time.

Bringing It to San Antonio

The work was introduced into the San Antonio area about 1924, by Dionísio. He was in the public art tradition, because he had worked on big public-art park projects in México City. When they came to San Antonio, most of the work they did here was similar, such as the work in Brackenridge Park and Alamo Heights. My dad did work for a company—I think the name of it was 'Pizzini Brothers.' They were in the business of creating these facades or doing the artificial stone concrete moldings and precise architectural details.

There's this legacy of my great-grandfather, my grandfather, my father, and now myself. We're talking about my grandfather being born in the 1870s or 1880s, and then my dad from around the turn of the century. Since my dad didn't have me until the late 1950s, we skipped a generation there. But a lot of things I know about concrete work and mold making are things I picked up from my father. It's the same with the 'trabajo rústico.' It's something that goes back to the 1860s and stayed popular through the 1920s. People kind of lost interest there for a while, but we continued to do the work. There wasn't as much commission work, like public art, then, but my great uncle continued to work pretty much until the early 1950s, until he became too sick to work. He passed away in his sixties, in 1955.

Influences of Father and Mother

I wish I would have known more of my dad [Maximo]*—what he was like when he was younger. In my formative years, my dad was in his sixties and seventies. My mom was quite a bit younger. She was just in her twenties and my dad was in his fifties when they got married. As far as I know, she was his first love.*

A 'faux bois' Torii gate-style entrance portal built by Carlos' great-uncle Dionísio Rodriguez for the Japanese Tea Garden maintained by Kimi Eizo Jingu in an abandoned limestone rock quarry. The walkways and bridges were built with prison labor. The site is designated as a Texas Civil Engineering Landmark, a Registered Texas Historic Landmark, and is listed on the National Register of Historic Places. Below, the maker's signature carved into the concrete sculpture. (Don Kirk)

He didn't marry for a long time because he had a family to take care of. My grandmother (my dad's mom) passed away at a very young age. She was a schoolteacher in northern México and she passed away when I think my father was four years old. So, my grandfather remarried and had another family of, I think, another five children. Later on, when my grandfather also died at a young age—I think he was forty-two when he passed away—my father was the father figure. So, he brought the family north.

I wish my mom had more opportunities when she was younger, because she is a very smart person. She wanted to go to Texas Woman's University in north Texas, but being the oldest of seven children, she was needed at home. I am eternally grateful for the sacrifices she made for my brother and me. She had us at a young age and worked hard as a bookkeeper for H-E-B, a job with little chance for advancement, but a job she retired from after twenty-seven and a half years. I do think she passed on some talents to me genetically. And then my dad, he was a very creative person. He could solve a lot of problems in his own way. He saw quite a few things happen in his lifetime—from the 1920s through the '90s. In the 1920s, he had one of the first cars available. He had the opportunity to drive a modern car and see airplanes and see all the new changes, and he lived through two world wars. By then, he had been married to my mom for thirty-seven years. My mom didn't remarry; she's taking care of my grandmother.

Confluence of Work and Family

My father met my mother and her family through his friendship with Dionísio. The two were always friends and my dad married my mother about three years after he passed away. Even with Dionísio gone, my dad continued doing the work, but on a much smaller scale, such as making garden furniture. I think it's very important to realize that the work my father did with the garden furniture created a whole new genre of 'faux bois.'

We still do it the way I learned it from my father and the way he learned it from his wife's uncle. It's been a tradition that's been passed down and I'm not really sure how many people are still doing it the same way. My maternal grandfather was an assistant to my great-uncle, his 'compadre.' Even though he didn't continue the work and do it himself, he was an assistant and did a lot of the form work—the grunt work.

The Dionísio Rodriguez Entrance Portal, Japanese Tea Garden, San Antonio, Texas. (Don Kirk)

華
園

An old snapshot of Carlos' father, Maximo, sitting on one of his works of art. (Courtesy Carlos Cortés)

It runs through both sides of my family. If you add everyone's years into it, there are well over a hundred years of my family working in some sort of concrete construction.

THE WORK

Examples of the Work

San Antonio is rich in Carlos' family work, including the entrance to the Japanese Tea Gardens (a.k.a. the Sunken Gardens), and at the old trolley stop on Broadway and Patterson.

There's the work in Brackenridge Park: There are several 'palapa' type structures in the park area; they imitate thatched roof benches that are tree formed. There is a bridge near the Hildebrand entrance to the park. We restored it in 2001. There are pieces on the River Walk here in San Antonio and at the Witte Museum. There's a bus shelter in the Lavaca neighborhood and there's work in our newest mall in the Northwest part of town, The Shops at La Cantera. There are also a lot of private commissions scattered throughout the city, especially in Alamo Heights.

Recent notable examples of Carlos' work completed since the interview with him include picnic tables, benches, and signage at the Robert L.B. Tobin Park along Salado Creek between Loop 410 East and Austin Highway, and the large-scale Grotto along the San Antonio River Museum Reach expansion. The latter is his largest work to date.

People can see many examples of my work at studiocortes.com.

Signing One's Work

Carlos is aware of the need to distinguish the creations of different Cortés family members.

> *I try to sign and date everything I make so there is no confusion, but I wish my dad would have done that. However, I can just tell by just looking at the work, who did what from what era. My father's technique was different than my great-uncle's, and mine is different from my father's. To be honest with you, I don't know what people will say later on. I just hope they will still be enjoying it. I don't think I'm the only person in Texas who does the work. I might not even be the only one in San Antonio doing it.*

Carlos taking a very short break for a "photo op" at his outdoor workshop. (Tommy Adams)

'Faux bois' trolley stop created by Dionísio Rodriguez at Broadway and Patterson in Alamo Heights, Texas. Listed in the National Register of Historic Places. (Don Kirk)

TYPES, PLACEMENT AND DURABILITY OF WORK

Outdoor Work in Natural Settings

For the most part, it's all outdoor projects that I do. I would say we do maybe 95 percent outdoor pieces—benches, tables, and birdbaths, or landscaping types of structures, like planters. We will occasionally do a bridge, tree forms, and sometimes these sculptures are used indoors—end tables, console tables, or even dining-room tables. I guess I prefer to see the pieces in natural settings, but I've also seen them indoors and they look nice.

Weather Effects on Outdoor Pieces

Carlos says weather can have adverse effects on the pieces.

If it's in an area where there's a lot of cold precipitation—freezing precipitation—where you're going to go through a freeze-thaw cycle where it rains and then it freezes and then it thaws again, there can be serious problems. What happens is, the concrete will absorb that moisture and since these pieces are made or finished in layers, any moisture between the layers will cause the piece to 'spall' and you can lose some of the detail. 'Spalling' means that it flakes off. Surprisingly, some of my pieces in Maine are holding up pretty well. I think it has to do with the fact that they're on the coast and that it doesn't freeze as much.

In the South—I would say from California to Florida and in México, these pieces can last practically forever. I've got pieces that are over eighty years old that are in very good condition. But, you take them north and they can start to fall apart within one or two seasons.

Therefore, the work is more prevalent in the South, but, he says,

There's plenty of our work up north. They traveled quite a bit, my great-uncle and my maternal grandfather. People heard about them by word of mouth in the 1920s and later. There's quite a bit of work up in Michigan and Chicago, but it hasn't held up nearly as well.

A 2004 palapa bus shelter by Carlos Cortés in the Lavaca/King William neighborhood, San Antonio. (Don Kirk)

THE PROCESS AND STYLE

Overview of the Process

Carlos explains the process of working with a client from design to delivery and installation:

Normally I have them look at my website, or look at photographs of my work. I have a portfolio with me they can look through with pictures of the work I have done in the

past. If we don't have a design that you like already available for you to look at, I will sketch something out for you. Usually, however, when someone comes over here and they're looking for me, they've already done their homework and they're sort of familiar with the work. Then, if they have an idea, a specific sort of look that they want, we try to work with that. You know, if they need it to be a certain size or a certain style, with a back, or no back or an arm rest.

Carlos likes to sign his work, mainly to distinguish his art from his predecessor's. (Don Kirk)

Stick Built or Organic

Do they want it organic looking or do they want it to look stick built?

Stick built is like a park-bench style that looks like it's been put together with twigs. In the organic style, it could be like a fallen-tree sort of look, where it looks like it just fell over and you know someone propped something up to make it into a bench. Everything is still sort of man-made looking. This chair that I'm sitting in, it's meant to look like a hollowed out tree stump.

From Commission to Delivery

Then, depending on the cost of the piece, we'll take a portion of that in advance—usually a third or half down—and then we work on the piece, and when the piece is finished, we deliver it. We've made deliveries as far away as Maine and California, so depending on how soon they need them, they can wait around here for a while before we ship them. We try to give them a finish that looks a little worn—not quite so brand-new looking. It could be three days to a couple of weeks. Usually when something's finished, though, we like to get it to the customer as soon as possible.

DETAILS OF PROCESS: THE BEGINNING

The Materials

Materials are easily found. We use Portland cement. We use clean sand, mason's sand, which is finer. And then we use a sharp sand,

Carlos starts a project by building an armature of rebar or copper pipe and metal lath welded or tied together with soft steel wire. (Don Kirk)

River Walk trellis arbor by Carlos Cortés near the Tobin Center for the Performing Arts (Municipal Auditorium), San Antonio, Texas. Constructed in 2001, vines were intended to be grown on the tree-limb "latticework."

(Don Kirk)

which has small aggregates, but is bigger than the finer sand. And that's pretty much it. Plus, water and rebar. We use hardware cloth and metal, all used in construction work.

It Starts with an Armature

Once the design is decided upon, construction begins.

First, we make an armature out of lengths of rebar welded together. Then we apply hardware cloth and metal to the rebar figure. The armature is the structural frame and it's usually shaped into the form we want to create. You can tell by the shape pretty much what it's going to be when it's finished. You really need to think ahead and find out where you want to place a detail like a knot or a broken branch or some sort of crevice. You allow for that in your armature. Everything's reinforced here. We start from the inside and work out. The armature is filled in with a concrete mix.

I like to let the armature with the really rough form cure for anywhere from seven to twenty-eight days. After twenty-eight days it's reached its maximum hardness.

Then another layer of sand and cement is applied and allowed to harden for a couple of days, or maybe as long as a week or even longer, depending on the piece. The next step is to add a bonding agent and a sand and cement mix, sort of like stuccoing up a brown coat.

The finish layer follows immediately after that—of pure Portland cement—what they call 'neat' cement. This is where all your texturing is done, where all the detail is done.

TEXTURE AND COLOR

Creating the Textures

To me, it's not trying to imitate a certain type of wood; it's more just trying to make it look rustic and like real wood, basically. I try to capture some of the grain pattern that would appear naturally if you were to saw a piece of wood in two. I try to do that.

Natural, But Not Real

I don't really go into that much detail. There probably are some artists or sculptors who want something to look exactly like a Xerox copy of nature, but I don't think that's part of the whimsy of this sort of work—what is called in French 'faux bois.' I've seen some forms that were done at right angles and they're still interesting to look at. You know trees don't grow that way naturally, but the idea is to suggest nature's ways, not recreate it.

My family worked that way; they weren't trying to be super realistic. My great-uncle was so good at the organic forms and his armatures. He thought way ahead to leave space for crevasses, or cracks, or where the bark was peeling away. He thought about all of this ahead of time so that when he was doing the detail, he had already allowed space to create them.

When I look at landscapes, the ones that I like are the ones that suggest that there is a flower there, but it could be just a touch of color. It's the same thing with this work: you do a little detail work, you scratch a little area with a screen or a fork or a trowel to suggest something you know would be there. Maybe the bark was peeled away. It doesn't have to be exact; it doesn't have to be perfect. Sometimes they are fooled, and when they tell me, it's a good feeling to know that I was able to do that. I could spend literally hours on details, but then it's no fun and you want to go on to the next project.

Carlos Cortés working on the finishing details of a garden bench. (Monte Adams)

Staining, the Final Step

The color is long lasting, but it doesn't go very deep. It stains the concrete, but only on the surface. Color is added after the piece has been allowed to cure maybe a couple of days. In most cases, we sand it down one or two days after that. Then, we apply a slip of the same sort of cement watered down over the entire structure. We allow that to dry and then we apply the mineral salt stain to that. We usually try to do it the very next day when the cement is still sort of curing and absorbing moisture so it will draw the color in. Other times, there will be some sort of highlights that are done in a fresco sort of way that's applied when the cement is still wet.

Making the Stains

I make the stains myself—cheaper that way, learned it from my father. The mineral salts are part of the secret. There's iron, there's sulfur, there's copper. You can find ready-made stains for concrete, but the differences between those and the ones I make are not that great. They're very similar even though they have a color chart with twelve different variations. When you take it home, it seems like maybe only three or four variations, not as many as twelve.

Detail of one of Carlos' benches showing the texturing done to make the concrete look like tree bark, Constance & Crawford Pocket Park, San Antonio. (Don Kirk)

You'll have an amber color, dark colors like a black; you might have a greenish blue; you might have a gray; then there's maybe a rust color. There aren't very many, but you can kind of play with those colors to vary it a little.

DETAILS OF THE PROCESS / THE INSIDE STORY

Is There a Secret to the Process?

People seem to think that our 'secret' is probably the staining. I want to be honest with you—there are no secrets. Everything has been covered before; everything's been written about. There are even books out there on the process of making the stains. There's plenty of information out there if people just look for it. It's true they have been sort of lost over time, but they are out there where someone could find them. There are books out there that were written in the 1940s that spell out everything—all the details—and they even show you the little tools to use. Sometimes people don't want to do that research and they want to just skip a few steps, and that's what I don't like. I learned it the hard way. I learned it next to my father and that's probably the best way to learn it.

Carlos' tool box with the necessary soft drink to keep him going. (Monte Adams)

Basic tools: paintbrushes, wire brushes, window screen, forks, palette knives—anything that can carve into the malleable cement before it sets up. (Monte Adams)

The Tools

Unfortunately, I don't have any tools left of my dad's. I use little pallet knifes. I use a bent fork to texture, to give it a sort of bark finish. A piece of screen—regular window screen—to do the graining. And a paintbrush and whiskbroom. I use the brush to apply the finish cement on to the texture. It's the last step. I also use the brush to sort of sculpt and move the cement around.

The Emotion of Sculpting

Carlos relates the emotion of sculpting:

With me, if it's the right feeling—I sort of know when it's right—I feel like I'm sketching with this tool. I can sketch a tree on paper, but when I can take the tool and run it through the finished cement, I feel like I'm sketching it instead of actually sculpting it. It's kind of a weird feeling, but that's the only way that I can describe it. It's a good feeling to see it coming together. And it's not something that just happens. It happens in stages. You have to begin it with some sort of scratch on the finish coat and then it starts to come together. You add little details as it starts to set.

H-E-B Treehouse sculpted by Carlos Cortés.
Witte Museum, San Antonio, Texas.

(Don Kirk)

A Good Assistant is a Must

Concrete sculpting is not a solitary art.

> *It's important to have a good assistant. You can physically do it alone. I've done it in the past, and I watched my father work by himself. But then I was his assistant. I was his extra hands. But it can be done. What's hard having an assistant is releasing some of the responsibility to someone else, but you have to make it work. I make sure I'm involved in the design part of it and oversee it to make sure all the steps are done correctly. I do the finish work—I want to make sure whatever leaves here is not going to come back.*

'Faux bois' tree inside the San Antonio Children's Museum. (Don Kirk)

COMPLETION

Weight and Delivery

> *If someone's interested in this art form, they need to know it's hard work; it's heavy work. Even a small piece can weigh a considerable amount. A small bench could weigh 100 pounds; the chair that I'm sitting on could weigh 200 pounds. For delivering the sculptures, I have a flatbed truck and a trailer. I've got a lift on my truck that I use as a hoist and the same thing on my trailer. Sometimes it's just a couple of guys moving things. So it works, but it's tough on your back.*

AN ARTIST'S EDUCATION

Carlos' education in this unique art was a natural, organic process.

> *I grew up around the work. My dad worked at home so I had the opportunity to watch him create ever since I was very little. And I learned it really through a sort of osmosis because even though I helped him, I didn't really think that much about it at the time.*

Our Lady of Guadalupe cast from a mold. (Don Kirk)

When I was eighteen, I thought I wanted to be a doctor. My goal was to go through school to become a physician, but at some point I went totally in a different direction. I think it had to do with my father needing my help. I was very confused in my early twenties, so I wasn't really sure of what I wanted to do. And later I met my wife, who encouraged me to continue the legacy that my family had started.

When I started doing this for the very first time—and I was doing these projects alone—I used to think about the beginning of a piece for so long that I would have to, sort of, will myself to get started. I had to think about the steps that I had to do and I was afraid of maybe failing and having somebody not like my work. But I've come to realize that at some point you just have to get started. Then once you get started, it becomes easier. And then the next project is even easier and then, before you know it, it just kind of flows. We are trying to create something that looks natural. It seems to grow—in layers. It is sort of organic—evolves on its own. I really don't call myself an artist; I have learned this technique. My dad didn't really call himself an artist either, but I think he was.

Making Molds

This is sort of a sideline for me, but my dad learned how to make molds at a young age. He was a mold maker, a pattern maker, and he made his own originals. I've got molds for flamingos; I've got molds for cranes; I've got molds for armadillos, for squirrels. He was very good at sculpting the originals. They look very lifelike and people still enjoy them. There are pieces that are over fifty years old or even older. We use the molds occasionally to make a birdbath or to make the flamingos as statues. Some of the older statuary pieces are very desirable; they're very sought after, espe-

BENCHES

A Carlos' 'faux bois' concrete park bench, this one at Carlos' studio. (Don Kirk)

Robert L.B. Tobin Park bench (Michael J. Smith)

A place to rest a moment from a stroll on the San Antonio River Walk. (Monte Adams)

cially in the antique concrete statuary market. Making art pieces with molds is totally different from what my great-uncle did. He was more into the free-form sculpting of the concrete.

Customers

Have you ever had anyone who was unhappy with your work?

I've never really had an unsatisfied customer. Other than with pieces that traveled too far north and are not protected. There's really nothing I can do about that. It doesn't matter what it is, Mother Nature will eventually break everything down. Mother Nature always wins.

One of his most famous customers was Martha Stewart, who featured him on one of her television shows.

She was down here; I think it's been about three years now. She saw one of my pieces in a garden show in New York City. It was purchased by a dealer here in San Antonio and was taken up there for a show, where she saw it, and became interested.

Sculpted concrete entrance sign by Carlos, Robert L.B. Tobin Park, San Antonio, Texas. (Michael J. Smith)

HIS WORK LIFE

Love of Work

I've been doing this for well over twenty years—actually most of my life—but as a business for twenty years. My dad never really put a whole lot of value on the work, at least outwardly, other than he said it was a way to raise a family, make a living.

I've learned to appreciate his work and my great uncle's work so it meant more than that to me from the beginning, but they were very humble about it. It can be very hard work, but the fun part of the task can make it easy. I think it kept my father

One of Carlos' creations, a picnic table, in Robert L.B. Tobin Park. (Michael J. Smith)

young because he worked until he was in his late eighties and early nineties. It gave him something to do, a reason for living.

I'm very happy. But, I'm not fulfilled yet. I think there's still much more work to be done. I would love to paint. I know that if I continue to do this, I will be happy. But, I also know that I won't have fulfilled my dream of being a painter. So, at some point, I have to do that also. I don't think I'll ever completely retire from sculpting; I'll just slow down. I love working with my hands. I don't know if I love being creative or I

A park bench in the same park. (Michael J. Smith)

THE SHOPS AT LA CANTERA

Carlos' work at The Shops at La Cantera, San Antonio, Texas. A slice of a very large sequoia-like tree for a tabletop, a dozen tree stumps for seating, and a tall torchère for a centerpiece make this "trabajo rústico" an attractive, functional piece of art. (Don Kirk Photos)

THE SHOPS AT LA CANTERA

ART FROM CARLOS CORTÉS

A bench growing out of a tree stump?

A mask that appears to be carved from a tree trunk is actually a playhouse for the kids.

Detail of 'faux bois' tree bark. How was it carved?

do it just to be active. If I'm sitting still too long, then I get edgy or nervous. I feel like there's something wrong if I'm not working or doing something. If I'm not working here, I'm working at home. I might be working in my yard.

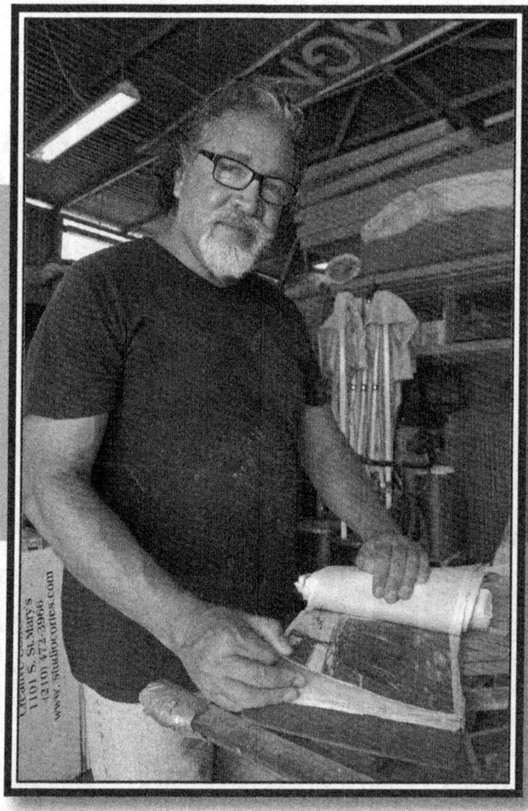

"I keep this clipboard of photos and idea sketches of mine for inspiration." —Carlos (Don Kirk)

I'm the oldest, though there are only two of us. My younger brother, a year and a half younger than I am, is now an Assistant City Manager in Seguin. He sits at a desk and he's more of a 'Type A' personality, I think, but he is also pretty active. It's hard for him to relax, too. My wife is always telling me I need to relax more, but that's just the way I am.

Taking Vacations

Vacation, what vacation? His work dominates his life.

The first vacation I've actually taken since I started the business was a year ago when I took my family to Disney World. I hadn't taken a vacation in over twenty years, I guess since the early '80s. There were little trips maybe on a weekend, but I don't call that a vacation. A real vacation where you're actually gone for five to ten days—not since the 1980s! It's hard to take a vacation when you're self-employed.

Never Quite Reaching Perfection

I wish that I could finish the pieces as well as I see them. And I can't draw them the way I see them, and I can't finish them quite

BRACKENRIDGE PARK TRELLIS BRIDGE, SAN ANTONIO
A MASSIVE PROJECT BY DIONÍSIO RODRIGUEZ AND MAXIMO CORTÉS, SCULPTED CONCRETE DESIGNED TO IMITATE NATURE'S TREES. (Don Kirk)

(Michael J. Smith)

LANDA BRANCH LIBRARY VALERO PAVILION
SCULPTED BY CARLOS CORTÉS IN 2008, SAN ANTONIO, TEXAS.

the way I see them. I see them in my mind's eyes. I keep trying and that's all I can do. I try to do the best that I can every time I do a piece and hope that people will like it.

PAST, PRESENT AND FUTURE

Restoring His Forbears' Work

I have worked on several different pieces that were done by my family. One in particular is one of the earliest pieces of their work—the trellis bridge in Brackenridge Park that was built around 1924, '25 by my great-uncle and my dad. It was restored in the 1960s by my father and then restored again by myself in 2002. It was really a funny feeling to know that I was working on something that their hands had worked on. And a good feeling to know that it was going to be around for much longer after I'm gone because we've taken the time to restore it.

He explains the process of its restoration:

We try to get to the armature so we can tie to it. We go down to the rebar and we have to assess the bridge—to see what was original, what had been

Simple idea sketches lead to fantastical works of art—this, a birdbath. Below, Carlos' "Project Idea File," a clipboard crammed full of idea sketches. (Don Kirk)

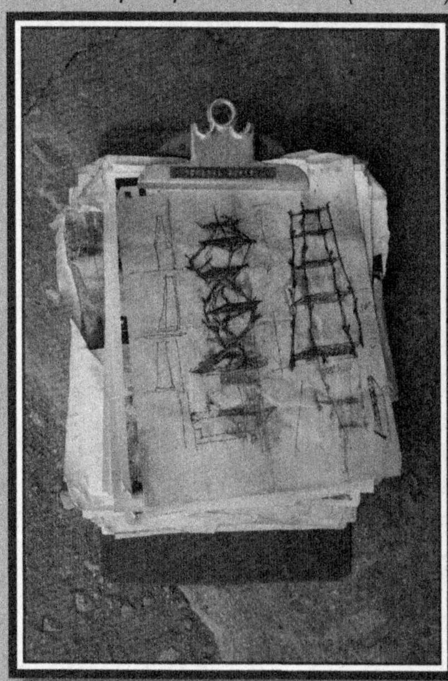

done as a repair by someone else. There was quite a bit of repair work done on that particular bridge.

He makes it clear that it is not a fast process:

When you do restoration work it takes longer. We have to remove whatever isn't original and start over. We get down to the rebar and we treat it for rust. Then we go from there and build the finish back up. A lot of the damage that was done to the original bridge was done because of vehicle traffic through there and maybe a couple of floods. And some of the big Public Works trucks that went by there damaged it physically. A lot of it was just done through careless damage. I think if we protect it, it could last forever.

He notes that some of his work is preserving a bit of history:

I've worked on some of the structures that are now declared 'Nationally Historic 'and 'Significantly Historic'—like work at the Old Mill in north Little Rock, Arkansas.

I Have a Dream

I have a dream to learn how to paint landscapes. I love landscapes; I love nature; I love trees. I love everything that's natural and I've become a fan of early Texas landscape painters like John T. Hill, Onderdonk, and Salinas. I have a dream to paint that way, and sometimes when I'm sculpting with my hands it kind of reminds me of painting with a paintbrush. I use a brush to do the details and I think it would be a similar sort of feeling to paint.

Look for Carlos' signature on his work. (Don Kirk)

He acknowledges that most of his work now is on a monumental scale—heavy, big, permanent, outdoor oriented. Painting would be much more intimate and indoors.

I'm thinking of the future when I can't physically do the heavy work. I am forty-eight and this requires a lot of physical work. I don't know how long I'll be able to continue to do that. My dad continued to do small-scale pieces until the age of eighty-eight or eighty-nine. For myself, I'm definitely sure the desire will be there. I just don't know if physically I'll be able to do it.

The Next Generation

I have three children. I have an older son who is twenty; the youngest is thirteen; and my daughter is fourteen. It's very possible that one of them will follow in my footsteps, but then again they might do their own thing. Even the oldest one is doing his own thing right now, but I had a late start too. Even though I grew up around the work, I didn't really take it up as a full-time job until my late twenties.

Carlos meticulously applying another of many layers of texturing. (Don Kirk)

The Future of *Faux Bois*

Despite the uncertainly of a family member taking over his business, Carlos believes the future of *faux bois* is secure—for now.

I think there are plenty of people who are interested in continuing this art form. There are people who were influenced in the 1920s, '30s, '40s, and '50s by my great-uncle's work; people who have been influenced by my father's work; and I think that people will be influenced by the work that we are doing now. There are mentions in magazines about this technique, so I don't think it will ever die.

HOW TO LEARN THIS CRAFT

Carlos doesn't believe this craft is currently being taught in art schools.

People think this is a lost art. But it's not lost; it's just forgotten. I think it just has to be in you to want to be creative and work with your hands. It's hard work getting started; it's heavy work—dirty work sometimes. You have to work in the elements—a lot of outside work. Actually, it's enjoyable, when it's not too hot to be outdoors.

FUNCTIONAL ART
BY CARLOS CORTÉS

"Give it a shot, give it a try. It's hard work, but everything—and I mean anything and everything—in life that you want to do your best at is going to take some work." —Carlos Cortés

Tree roots become benches on Carlos Cortés' "Mushroom" Palapa, a bus stop at the corner of Newell and Camden, San Antonio, Texas. (Don Kirk)

"When it comes to anything of metal, stone, or wood, it just intrigues me because a true craftsman doesn't just follow his trade, he follows others' trades." —Carlos Cortés

Carlos Cortés' whimsical Museum Reach River Walk Grotto, San Antonio, Tx.

RIVER WALK GROTTO

RIVER WALK GROTTO BY CORTÉS.

Alien gargoyles from the deep? Carlos can tell you what every part of this wonderment symbolizes; nothing is really at random, especially true art.

Dare you enter?

(Don Kirk Photos)

Medieval half-timbered construction inspired this mailbox by Carlos. (Don Kirk)

There is one necessary ingredient for success.

You have to have a desire to be creative. I think it's in a lot of us to be that way.

School Subjects that Help

Math is very helpful; a life-drawing class or a drafting class could definitely help. I've always loved to sketch and when I was in elementary school I won a couple of art competitions in school. In high school, I took some drafting courses. I actually worked as a draftsman for a while. All these basics have helped me in everything I do here. Everything that I've learned I've put to good use. Whether it's mathematics or some sort of structural steel application or just common sense. Botany could help, to know trees, how they form—know it from the inside out.

Getting Started

You can get started on something like this by going to the library and reading some books on sculpting with concrete. The 'How To' books are out there. There are some books that were written in the '40s about how to sculpt with concrete and do this 'faux bois' woodwork; they're just out of print. They'll teach you how to do armatures, how to make the mixes, and how to apply it to the armature. The rest of it is really up to you. I mean, it's practice. I had to do the same thing; I had to practice. It just didn't happen overnight. I encourage everyone who asks me to give it a try and to experiment on his or her own. It's sort of sculpting, but your medium, the cement, sets up a little bit faster. And, if you make a mistake, you can always kind of chip away at it the next day. Also, being that it's done in layers and sections, you can add to it and change it. It's not like marble, which is unforgiving.

My best advice is to learn the basics: how to create, how to build armatures. Then just a little experimenting, which is how you learn. Give it a shot, give it a try. It's hard work. But everything—and I mean anything and everything—in life that you want to do your best at is going to take some work.

Apprenticeship

You really have to spend some time apprenticing, some time doing the dirty work like unloading the truck or unloading the materials. I think you have to spend time hanging around with the person, just observing him to see if it's something you really want to do. The desire has to be there.

The Work Ethic

I want to do the best work that I can do, but that's in everything. From the time I was little I wanted to do my best. If it was sweeping a floor, I wanted to make sure that I did it right. I think it's genetic. I remember when I was twelve—selling Christmas trees at an H-E-B when my mom was working there as a bookkeeper.

A garden bridge in concrete—only a babbling brook is needed. (Don Kirk)

Carlos received the commission in 2013 to replace a rotting log-rail fence at the Gristmill with this 'faux bois' creation which should last much longer than the original fence, Mission San José, San Antonio. Below: One of Carlos' 'trees' on the San Antonio River Walk contrasting with the 'modern' Tobin Center for the Performing Arts, designed by Marmon Mok Architecture. (Don Kirk)

I wanted to do the best work that I could and sell as many Christmas trees as possible. Or, if I was stocking shelves, making sure they were all lined up just right. I think you're just born that way. I've thought about that before—whether you're born that way or you're raised that way. I think a lot of times you can raise your kids the same exact way and it's just the way that they are made, what's inside them. I think our kids are a little spoiled now, but out of those spoiled kids you're going to get some good ones too.

THE ARTIST AND HIS ARTISTRY

If It's Not Fun . . .

My great-uncle was very good at it; he did some really whimsical things—like putting faces in the bark details and creating these shapes that looked like something else, some like human forms. I think you should have fun, and I think if it isn't fun anymore, then I would probably move on to something else. I don't want it to feel so much like work, though it is, of course, physical work. But physical work can still be fun, and when you see something go from nothing to a finished piece, then it's a good feeling. I guess I feel even better once it's finished and the customer likes it.

It's Got to Function

I want to have fun and yet I'm very critical of things all the time. If there's something that isn't the way I sketched it out or if we finish a detail that isn't comfortable, like a bench that isn't comfortable, then I feel like we kind of cheated the customer. You know, it has to be comfortable—more than if I were doing it for myself. If someone is paying me a lot of money to finish something out, I want it to be comfortable and beautiful. I want them to like it. That way, it gets me past that worry about it being perfect. Comfortable yes, and all that work that you put into it needs to be there, but the view is so subject to interpretation.

> **"I think our kids are a little spoiled now, but out of those spoiled kids you're going to get some good ones too."**
> **—Carlos Cortés**

A sampling of Carlos' religious works demonstrates his sculpting skills in concrete without using trees for inspiration.

A wonderfully detailed face in concrete; a far cry from 'foux bois.' This is "Jesus in the Garden of Agony," at Our Lady of Mount Carmel Church, San Antonio, Texas.

'Juan Diego' statue at the St. Matthew Catholic Church Sports Complex, San Antonio, Texas.

(Don Kirk Photos)

Statue of 'St. Isidore, Patron Saint of the Farmers,' Annunciation of the Blessed Virgin Mary Church, St. Hedwig, Texas.

Carlos' concrete cross at Annunciation of the Blessed Virgin Mary Church, St. Hedwig, Texas. 'Totus Tuus' was Pope John Paul II's apostolic motto. It's a Latin phrase meaning "totally thine." (Don Kirk)

Birdhouse stuck atop a tree branch and one of the 'Stations of the Cross' at the Oblate School of Theology. Below, a 'foux bois' driveway-gate arbor. (Don Kirk)

Carlos tries to sign his work whenever he can—most not so boldly as this—to distinguish his work from that of his predecessor and others, though he'll tell you, each artist has his own unique style one can identify if he looks close enough. (Don Kirk)

Use of the Sledgehammer

Carlos has guidelines for living and working that have served him well.

> *Dad used to always tell me, 'Don't worry about what someone else thinks, just do what you feel you need to do and there is no one who is going to tell you that it's right or wrong. So, that's what I try to do. When it comes to this sort of work, some people think that I'm the expert. I'm not the expert. To be honest with you, I have doubts, second thoughts about my work lots of times. I will take a sledgehammer to something if I don't like it. For example, if there's a bench and maybe the back is angled just a little bit off and it leans back too much, I don't have a problem busting it up and starting over.*

Expertise and Individual Artistry

Carlos acknowledges that the real expertise in this field comes through the sculpting, the staining, and the final texturing.

> *And some of it is in the actual design work and putting the armature together. I do think what differentiates my work from my dad's work—and ours from my great-uncle's work—is the finish. That's something we can't duplicate. Everybody's got his or her own sort of finish details. I think I can tell better than anybody*

"When you see something go from nothing to a finished piece, then it's a good feeling."
—Carlos Cortés

(Michael J. Smith)

else, because I've just been around it for so long. You learn it all, but then you put your own individual stamp on it.

Other Styles

There are so many different styles of this work. I've seen some faux wood work that looks so realistic, and some of these are big companies that are doing this work. You know, like for Disney, for themselves—theme-park work. And then I've seen from the art side, which I think is also very good. It isn't as realistic, but the ideas that they come up with are very interesting to me. That's what's neat about this work—they can all be different, and it can have a lot of value to it. There's a lot of room for creativity in this world, in this work.

Lucky for us, we have been graced—for generations now—by the passion and dedication of the Cortés family through their artistry and unique *faux bois* creations. Feel free to stop by his outdoor workshop on St. Mary's Street to see what he's up to now.

Carlos' sculpted-in-concrete upturned fireman's helmet for the 'Fallen Firefighter Memorial' at the San Antonio Fire Training Academy. The helmet was done in collaboration with public artist Jesse Amado. (Don Kirk)

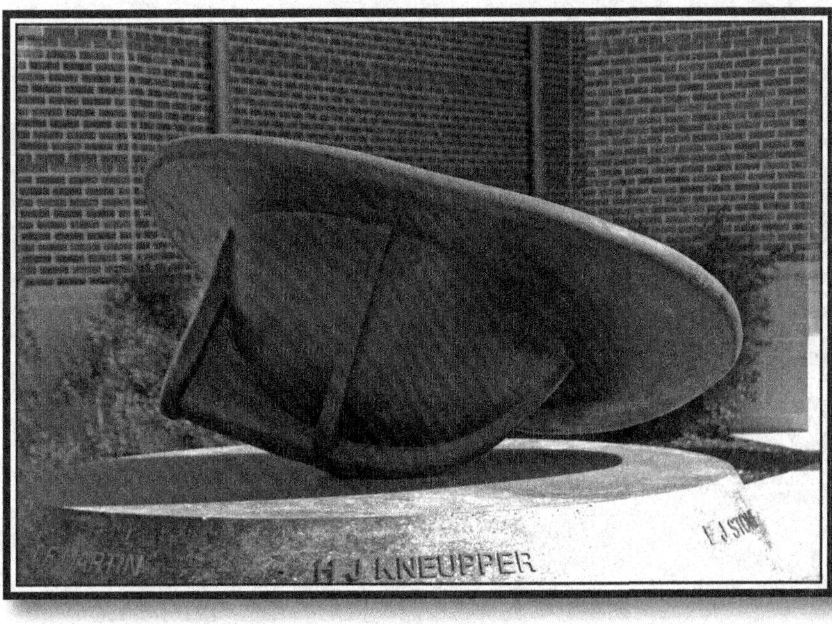

5. Othon Garcés

(Monte Adams)

INTRODUCTION

OTHON GARCES, owner of Garcés Metal Specialties, is another craftsman following in his forebears' footsteps. His grandfather was a blacksmith.

Many skills are required to be a successful metalworker, especially in the area of restoration. From drafting to building models, to casting, cutting, and welding, each challenge is unique. Projects may require custom designing, even forging. Tools can range from the simplest hammer to large, expensive cutting machines using the latest technology.

Othon's artistry can be seen throughout South Texas, gracing such treasures as the McNay Art Museum, the Southwest School of Art, the St. Anthony Hotel, the Alamo, and the Witte Museum. Othon's passion is transforming his vision into permanent metal constructions that transcend their sometimes pedestrian functions.

Othon is concerned about passing on his skills: "This is a trade that I wish there were other people, men or women, that I can pass it on to."

Othon Garcés.

Othon Garcés

METALWORKER

Othon has been a metalworker for more than sixty years. His father, Juan Garcés, learned the trade from his father in México. Juan came from San Luis Potosí, a mining town, and Othon's grandfather worked as a blacksmith.

> *He would work in the silver mines, in the blacksmith shop, repairing the pumps, the picks, the shovels, or whatever had to do with metal. So, my dad followed along in his footsteps, and so did I. My dad came to the United States in the 1950s. He just wanted a better education so came to the United States and went to school at Fox Tech, called the 'Main Avenue High School' back then. There he took drafting classes and what they called 'sheet metalwork.'*

Juan Garcés immigrated to San Antonio from México in 1915. He was a man of few words, but large by example. He taught his metalworking and artistic skills to his sons. One member of a family of nine children, Othon started helping out in his father's workshop when he was just six years old.

> *All he had to do was look at us, and we knew what he wanted us to do. I just loved to help him out and work alongside him.*

Othon points to a certificate on his wall that his father had received:

> *It's in Spanish. That was an automobile correspondence course in Los Angeles, California. So even though he knew auto mechanics, he chose to take the path of a blacksmith. My dad was an avid reader. He would tell us, 'You need to read, son, you need to read.'*

Drafting and Constructing

> *Of course my dad had to learn English—learn to read, write, and learn drafting. He loved to draft. He was a good draftsman. He felt that if you could draft it, you could make it. And his handwriting was excellent.*

Othon's father, Juan Garcés (on stair) and Bob Ingram (kneeling) installing a wrought-iron handrail in October of 1925. (Courtesy Othon Garcés)

Othon explains how drafting is critical to the metalworking process:

> *You start by taking the work you're going to produce and put it on paper. That's what you use as a model to make all your forms, the dimensions, and how it's going to be made. It's like an image that you implant into your mind. I'll always remember Dad saying, 'You should be able to lay it out and draft it.'*
>
> *He was a constructor. My dad was talented. He would do work that you would think it would take ten people to do, but he did it*

with two. It was only him and his helper. And then, as we grew up, of course, we were his little kids. Unpaid helpers. We were the ones that would take the sledgehammer to the metal. He would just point where he wanted us to hit and you better get out of the way.

Eventually, they did learn to hit their mark.

Accuracy and respect, and be here, and just do whatever. There really weren't too many words that he would say. He just looked at you and you better pick up the hammer, or pick up the file, or whatever he wanted because he'd just look in your direction and you better know what he wanted—the crescent wrench. You'd give him the wrench because you could see there was a bolt to tighten.

Growing Up

Othon remembers his father as a man of few words.

All we would do is just listen. You weren't allowed to sit around when the adults were talking. All they had to do is look at you and you better step outside, move on. That was just their way of bringing us up. We are a family of nine—three boys and six girls. We grew up here in San Antonio in the Five Points area, near San Pedro Park. My dad was very thrifty and wise—very wise.

We had two houses that we lived in; we just had too many people. Each was like a two-bedroom home and right by San Pedro Creek. One was for the boys and the other was for Mom and Dad, and the girls. The one where we lived had that shop in the back; it had a little garage where he worked. They don't exist anymore. When the IH-10 freeway was built, it went through some of that area and all of the houses were demolished. We moved to a house on West Dewey, right next to San Antonio College.

Today, the Othon Garcés family lives behind St. Mary's University.

I've lived there for forty years.

A Family of Independent Souls

Othon was not the only one to follow in the family business. Basically, all three boys did. Othon explains:

The other two brothers also have a business and do similar work like I do, but we just never did get to work together. Everybody's very independent.

Spiral staircase to the library. Railing & balusters by Garcés Designs, McNay Art Museum, San Antonio, Texas. (Don Kirk)

His father's company was Garcés Iron Works. Othon explains:

> *You know, I broke away from my brothers. My brothers have that business. I started my own company in 1979.*

Though his brothers do some restoration work, they don't compete.

> *Fortunately, we never cross paths as far as architects, contractors, and homeowners go. Everybody has his own little niche and there's plenty for everybody. Our dad passed away at eighty-six. He worked until the very end. When I would go visit him at home, he would always ask me, 'Did you order the materials for a certain project?' I would tell him, 'Yes, Dad, we have it ready. We're working on it. We need to do this or we need to do that.' And that was enough for him and then he was satisfied. I mean, he was very active.*
>
> *My mom passed away at ninety-six years old. We're all self-sufficient. You know, we're nine in the family, but really we're not a close family. Everybody does his own thing. We're all quite different.*
>
> *One of the twins is a retired school teacher. She worked for San Antonio Independent School District for thirty-three years. My oldest sister, Magdalena, worked for San Antonio College for forty-four years. She passed on; the rest are all still here. We're all kicking. Kind of slow, but we're kicking.*

Detail of McNay Art Museum library staircase railing. (Photo at left Courtesy Garcés Designs)

(Don Kirk Photo)

A Woman-Owned Business

The Garcés company's employee handbook has the history of the company, including the fact that his wife Natalie was the president at the time of this interview [Othon's daughter Melinda is now the president]. A woman-owned business has some advantages in getting work, he notes, but they have so much work it has not been necessary to tout that very often. Though he didn't really know Natalie growing up, she lived nextdoor to his father's shop. They "went around" for four years and finally got married when Othon was twenty-four years of age.

Candle chandelier by Othon in the chapel at Southwest School of Art. (Tommy Adams)

Design and Materials

I design or I help some of the architects or designers. If a client comes to us and wants a special design, I usually like to get pictures that I can follow, so that I can have an idea what they're looking for. Or, I go to their residence or place of business to get a feel for what the building should have. I use that as sort of a model not only for me to design, but to get them involved in the design phase of it. If it's a homeowner, I always say, 'You're going to live here, not me, so let's design something that you're going to like for a long time.'

We work with regular wrought iron, which is steel, and with brass, copper, stainless, pewter, and tin. There are different varieties of steel that you can use in our trade; it's not just one item.

Blacksmithing Versus Metalworking

To blacksmith, you take mostly all steel and you forge it to form it. For metalworking, you take different shapes that are already out there in the industry and you combine them to fabricate a product. Using sheets of raw steel or copper or stainless, you incorporate that into the design. Blacksmithing is really an art, since you have to form it. There's nothing that you buy off the shelf, you literally create the shape of the metal and form it into whatever you need.

Othon's wall candelabra sconce, Southwest School of Art. Below, a colonial light pendant. (Don Kirk)

An Ancient Craft

Metalworking goes back to the Egyptian era. The first metal used was copper. Blacksmiths and metalworkers made the tools for the masons.

Othon watched a documentary about blacksmiths on the Discovery Channel.

When it comes to anything of metal, stone, or wood, it just intrigues me because a true craftsman doesn't just follow his trade, he follows other trades. When they were making the pyramids, they would have a certain area for the blacksmith to work. The metalsmith made the points to carve out the bricks and the slabs of marble.

Global Nature of the Business

The blacksmithing trade has been around in not only the Mexican culture, but it's in the French, the German, some Italian cultures. The blacksmith is in all cultures, really. Look at pictures from Europe and you see examples throughout. Of course, a lot of those ideas today come through decorators and designers.

We have had clients take pictures of some of the chandeliers from either México or Europe, and we incorporate their ideas. You might say we copy some of the ideas, but we make them work to suit their conditions. Much of what you see in México was brought over by the Spaniards. Also, the French, and many Germans went into México.

Tools of the Trade

In this trade, of course, you use chisels, files, and the forge to get the metal hot. The tools that a blacksmith uses, they're basics. You have your hammers, the forge, and your tongs—I mean, those don't go away; they last forever. So, even though it's a very old trade, there are new cutting-edge tools that you use. Yes, even your

The old standbys for metal fabrication: chop saw and drill press. (Don Kirk)

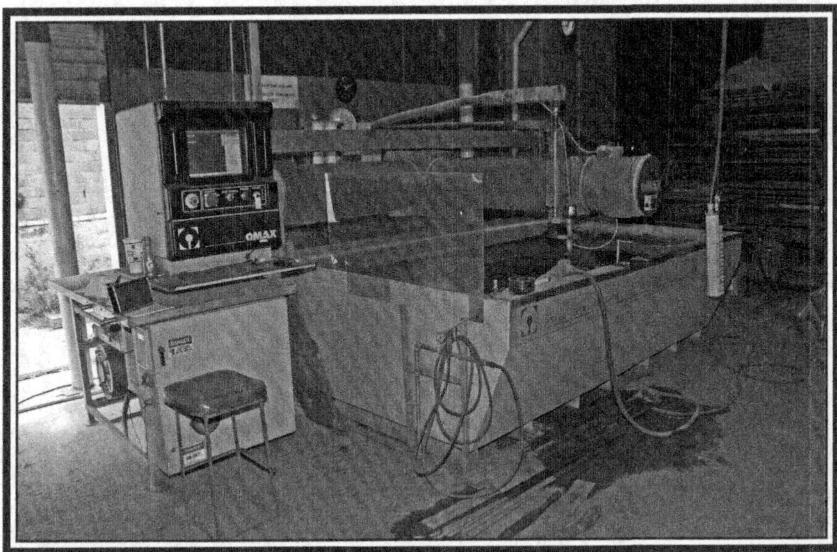

The OMAX 55100, a water-jet cutting machine using a garnet abrasive and high-pressure water. It pressurizes water up to 55,000 pounds per square inch and forces it through a small jet, creating smooth, clean cuts through a variety of materials, including metal, stone, marble, granite, ceramics, glass, wood, and rubber. (Don Kirk)

St. Anthony Hotel, wrought-iron railings, Garcés Iron Works.

Lobby staircase and balcony of the 1909 St. Anthony Hotel.
Sadly, it was removed in 2014 during a remodeling.
The hotel is listed in the National Register of Historic Places.
(Barbara Dean Hendricks)

Exterior balcony railings. (Don Kirk)

Detail of the elaborate wrought-iron scrollwork on the upstairs balcony of the St. Anthony. The work was hand forged by Othon and his father. (Don Kirk)

welding equipment, your welding machine—I mean they're a lot better; they're smaller and not as heavy and bulky as they used to be. It gives you a better product. Now we've modernized—we have a machine that's called a 'water-jet.' It's one of the state-of-the-art cutting machines that cuts with garnet and high-pressure water. It's fully computerized. It's one of the machines I saw that I needed to purchase to be able to stay in the market—to be able to bid on some of the more intricate projects. Not only is it good for the artistic work, but it also does very precise machine parts.

We have some new techniques of welding now, which used to be stick welding. The Heliarc welds aluminum, stainless, and brass using a helium and tungsten electrode. Now, it's not only the cutting part; now you have plasma. The welding industry has really improved in the past ten to fifteen years.

The Process: Bidding

On most, you have to bid on them first. This is where the know-how of what it's going to take to do a project comes in. The other thing is that they have to either know you or know enough about you—that you're capable of doing it. What has kind of muddied the waters in our trade is people coming into the trade thinking that if so and so can do it, I can do it—until they start to do it.

Othon hooked to a safety line as he welds ironwork at The Bushnell in San Antonio, Texas. (Tommy Adams)

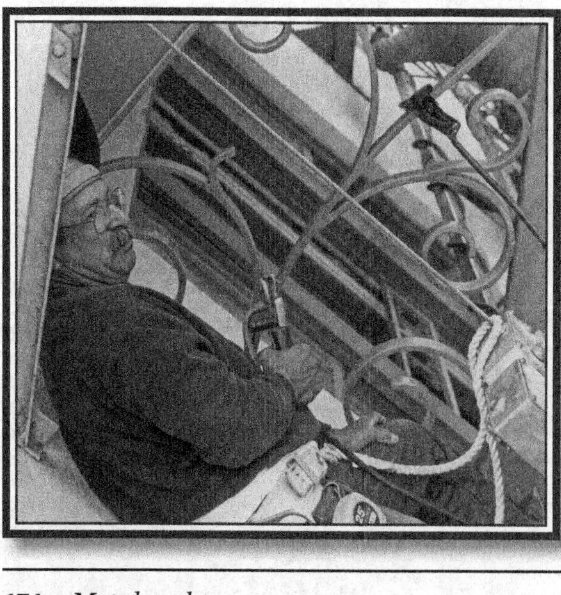

The real key is knowing how to bid the thing—knowing what it really costs. That takes years to figure. You have to learn your trade, know it well enough to do that. That comes like a gift. It's hard to say when it's going to take an 'X' amount of hours. Sometimes I have heard I'm too expensive, but once I total up everything I come

very close to what it should have been. And, when I go down a little bit, I say, 'Oh, I should have kept it at the price I had it.'

Installing large balconies on The Bushnell, a seven-story 1926 Mediterranean-style apartment building/condominium in San Antonio. (Tommy Adams)

Finally, you win a bid; then you get your materials—your drawings—you have to draw up working drawings. Once those are approved, then you start manufacturing. When everything is manufactured, you install it. This is one of the things I do with whatever project in which I get involved: I not only want to make it; I want to install it. I just feel that I don't trust somebody else putting up my work the way I do it. I think they might mess it up or break something or screw it in wrong, or it won't fit, you know.

Materials and Relationships

Most of the materials are bought here in San Antonio, either at Alamo Iron Works, Triple-S Steel, or *Reliance Metal Center. A lot of our copper we get from Farmer's Copper here in San Antonio, and there's Westbrook Metals.*

I'm very particular about specifications. I know what I'm going to use, and I want to be the one to get what I'm going to use in my project. I'm pretty good at estimating the amount of material I'm going to use. I usually don't have scrap left over. I learned from my dad—be thrifty. You have to have very good resources, not only about the materials, but who's behind the counter.

The restored 1896 Romanesque Revival-style DeWitt County Courthouse, Cuero, Texas, showing Othon's custom roof-vent finials and tower corner finials.

(Don Kirk)

Once I know them, I go and usually talk to just one salesperson most of the time. That way I know he's telling me the truth—that he has it or doesn't have it, or he tells me where I can find it. And we have a good relationship. I've got some salespeople who we've been dealing with for thirty years. I have one gentleman, Johnny Brenham from Reliance Metal Center, he has been in the trade forty-plus years, and we still deal with each other. So, there's still a lot of us around.

The DeWitt County Courthouse: A Restoration Challenge

The courthouse in Cuero was built in the 1800s and remodeled in the 1950s. Othon relates some background:

In the '50s, all of these staircases were demolished; the openings were closed over and renovated to look like the '50's. There were four staircases going from the first to the second floor and from the second to the third floor—four different staircases. A few years ago we were called on to restore these stairs by the Williams Company. They're a restoration architect firm out of Austin. The contractor was J.T. Michaels.

We are registered restoration blacksmiths and metalworkers in the State of Texas, so they already knew about us through other

Detail of Othon's fabricated-in-metal finials, DeWitt County Courthouse. (Don Kirk)

Othon recreated the three-story iron steps and risers, hand-forged railings, and balusters at the Dewit County Courthouse, Cuero, Texas. (Don Kirk)

projects. This project started off from scratch—based off a photograph. It was a real challenge. When the staircases were demolished, they really were! Possibly even melted. Even the pieces couldn't be found.

We went to Hallettsville. The Lavaca County Courthouse was built at about the same time, and it's in an area close to Cuero. We took photographs and dimensions and took some designs from that—along with the architect—to come up with the present design.

As of this date, we have all of the four staircases in place. Right now we are working on the handrailings. When we do a project like this, we work some with drawings, but this is a bit backwards.

What I do is make a wood mock-up. I work with some cabinet shops to help us to make a mock-up of three or four stairs and from that we kind of get the architect's input. Plus, not only the project architect, but there's also an architect from the Texas Historical Commission involved in approving all of the pieces that we're putting into the courthouse.

Then, of course, I'm fortunate they trust my input into how I would have designed it, or how would a craftsman design it. I mean you might draw it one way, but they wouldn't have actually made it that way, so we come to a compromise. I'm fortunate that the architect accepts my input. Then, of course, we pass it through the engineer to be structurally sound, but normally these projects are through drawings.

A Subtle Art

Metalworking takes keen insight. One can analyze something, but without experience, one cannot fully understand how a piece should be made.

I had a friend of mine say we are artists because we work with the heart, from the heart. When you're an artist, it just seems like you just have to do everything yourself.

Othon explains how he used the one photograph to rebuild the stairs at the DeWitt County Courthouse:

We enlarged it—like ten or twelve different sections. Then, we pieced it together to get an idea of what to do—not only for myself, but so that I can show my employees what the finished product is going to look like. They don't want to go out there if they don't even know what they're doing. I always keep them in tune as to what it's for, where it's for, and this is how what you're making will be used.

Sometimes Othon works only from photographs—as he did on the Cuero courthouse job. There, Othon and his workers meticulously recreated and installed three floors of iron staircases that had been ripped out and destroyed in a "modernization" remodeling in the 1950s. (Don Kirk)

A Garcés-fabricated steel column in the Dewit County Courthouse, Cuero, Texas. (Courtesy Garcés Designs)

Pre-Visualization

A lot is visualization, and, of course, we go with pre-existing conditions and then we have to make it work. We're not going to move the floors to put our staircase in there; everything has to fit.

Timelines

The first phase of the DeWitt project was the exterior renovation. We were involved in making all of the pinwheels on the outside of the courthouse. Here again, there was a project that we worked with nothing but photos. For dimensions, the architect had fun. All of us did. We figured them out and got them to look proportional. Because a picture sometimes shortens that. It can be very deceptive.

Othon still enjoys welding in his workshop on Zarzamora Street. (Don Kirk)

The project turned into a four-year effort, moving inside to do metalwork and some courtroom benches. Othon feels confident his work will last.

Unless they demolish them, they're going to stay there. The building has been there since the 1800s, and if they had left the stairs in there they would still be there. There's nothing to deteriorate. You should see some of the courthouses throughout Texas; they still have these wrought-iron stairs.

"A lot is visualization, and, of course, we go with pre-existing conditions and then we have to make it work. We're not going to move the floors to put our staircase in there; everything has to fit."
—Othon Garcés

Stable stalls converted to cafe booths with Othon's craftsmanship, Sullivan Carriage House, San Antonio Botanical Garden. (Don Kirk)

THE SULLIVAN CARRIAGE HOUSE

Gift shop gate detail, Sullivan Carriage House, Botanical Garden. (Don Kirk)

Other High-Profile Projects

Othon's company worked on the Sullivan Carriage House project at the Botanical Garden, along with Curtis Hunt and his company.

> *There are some stalls in there for the horses and we worked on that with Ruth Johnson from the Southwest Craft Center* [now Southwest School of Art]. *That was one of her projects. We converted the stalls for the horses into little booths for the restaurant area, plus we made the grills for the gift shop there. There were also grills on the windows. Some were already there; we duplicated the one on the north side using pictures we took.*
>
> *Here again, this is where the artist comes into play: You have a window opening; then you make it fit. Plus, you have to make it look like the photograph. That's where artistry and the craftsmanship come into play. It's all I know how to do! I have fun at it.*

"I had a friend of mine say we are artists because we work with the heart, from the heart. When you're an artist, it just seems like you just have to do everything yourself.
—Othon Garcés

Historic Homes

Othon works for residential clients as well as commercial work. He did work for the Tom Benson family and the Curtis Gunn family, and he worked on the Brown Estate. He appreciates the attention paid to historical structures in San Antonio.

> *I think we have been fortunate to have people like the Conservation Society keeping some of these houses even though they might look like they're not worth anything because they haven't been maintained. Sometimes organizations or foundations take them over and they're restored. They start from the foundation and work on up—the wood, the masonry, the metal, the painting, and the roofs. You see all of this craftsmanship from all of these other people, especially in certain areas of San Antonio.*
>
> *We worked on the Steves Homestead in the King William area. There's another house across from the Pioneer Flour Mill; it used to be the Patterson's house. We did all the metalwork at that house. We also worked at the Guenther House—some of that ironwork in the restaurant area and the exterior.*

Othon's wrought-iron fencing, Charles Butt residence, King William Street. (Don Kirk)

For Charles Butt, we did the fencing along his property and also made the rear entrance to his house. There's a gate there. For the Arsenal, we did all of that fencing on the north side and the east and the west sides. We did all that work back in the mid-'80s.

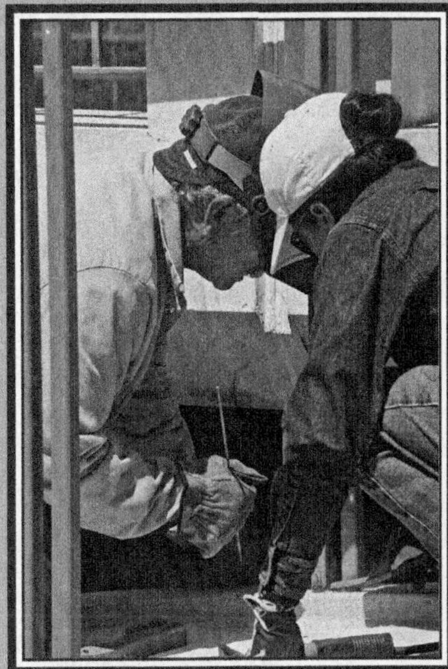

Othon and daughter Melinda arc-welding fencing into place. (Tommy Adams). Below, Othon's wrought-iron brackets and railings at the Guenther House. (Don Kirk)

Synergy in the Crafts

For this project at the Cuero courthouse, we've been working with Curtis Hunt also. Curtis Hunt is another one who has been around, I think, for generations. I don't know if you saw that article on the background of the Alamo they did on us, Curtis and myself. Curtis is a fourth-generation mason. I was a second-generation craftsman to work on the Alamo. [The reader will find Curtis Hunt's story in Chapter 6.]

The Alamo

Many of the craftsmen featured in this book worked on the Alamo. Othon forged some of the metal hinges for the Alamo restoration work and notes:

At the Alamo, we did the black stanchions on the inside, plus there were some supports to hold up the chains that keep the people from touching the walls, because the oils from the people's hands will deteriorate the stone.

Edward Steves Homestead, 509 King William Street, San Antonio. One of the San Antonio Conservation Society's fifteen properties. This one is open for tours. (Don Kirk)

Work on the Alamo marked Othon's entry into restoration.

About fifteen years ago, we started doing work at the Alamo—just some small projects. Then the people from the Alamo would recommend us. They'd be asked, 'Who do you use for your work?' and they'd say, 'Well, we use Garcés.' This is how I got into restoration work. Some of the contractors we've done work for—G.W. Mitchell, M.G. Boyle, Lyda, Guido—contractors that have been around for many, many years.

Future of His Business

Othon believes it is important to save his craft for future generations.

I was fortunate to carry out my dad's trade, but I have four girls and no boys. So, you can see I am probably the last of the tradition of being a metalworker. My daughters, they're in their thirties and forties. My daughter Melinda does now run the business.

It's a different era nowadays. This is more of a one-man operation. This is a trade that I wish there were enough other people—man or woman—that I could pass it on to. I've been looking to find somebody that is interested, to carry on the trade. I will teach anybody;

I can't take my skills with me. It has long hours and you might say the pay is not the best, but when you see something completed, that's the reward—that's the pay.

It's provided a living for my family; now it can provide a living for somebody else. We're going to have to recruit some people. Maybe this book will help recruit somebody. It's a skill you don't want to lose. I've even gone to high schools and tried to get some students. At the beginning I would tell them 'I don't want anybody who's an expert welder; all I want is somebody who can show up at eight in the morning and read and write.' You know, even if I have to teach them how to read or write, I will. 'To measure and follow directions, that's all you have to do. Just follow directions.' Sadly, I haven't found anybody.

Working with Clients

Othon had recently finished a project for a home in the resort area of Destin, Florida, at the time of this interview. A Turkish architect found Othon through a friend. The interior and exterior railings were stainless steel because of the potential for rust in that climate, and Othon's task was to produce a stainless steel handrailing of unique design.

Iron cresting lovingly restored by Othon on the Steves Homestead. (Don Kirk)

On the Novak residence in Elm Creek, the man and his wife would come to the shop and they'd design it together. Othon also did work on the Robertson's residence at Elm Creek. Othon discussed the importance of direct knowledge of the process when working with a client:

How do you report on the progress of manufacturing and installation if you're not involved in the project yourself? You have to be able to rely on your employer. Sometimes, if they know what they're doing, the project looks good. Other times,

The congenial Othon Garcés posing for a portrait. Below, welding a post to a base plate. (Tommy Adams Photos)

when I see it and think, 'Naw, he didn't get my message right,' and then I get involved. I say, 'No. we're going wrong here; we need to straighten it up; we need to change this or whatever.'

Twenty-First Century Tools and Ideas

Although much of his work is restoration, Othon also produces a good deal of new work. He sees a change in how clients get ideas:

In the last five to ten years, most of the clients in San Antonio already have an idea because of the Internet. They can pick up design ideas. Some architects can now draw so quickly with AutoCAD, I let them design unless it's a real good client who says, 'I want you to do the design.'

I'm not that computer knowledgeable. I think you get forced into learning the computers, either by looking at somebody else and how they use it or asking questions or reading. I read the manuals—as hard as it is sometimes to understand some of those manuals. I didn't know anything about computers, but now I can do some drawing on them; I can do some invoices and statements.

Study Courses to Take

I think one of the main courses to take if somebody wants to go into this trade is accounting. We might be good craftsmen, but we're poor business people when it comes to keeping the books. And if your bottom line doesn't look good, you're going to go under. Also, your math, your reading. I also think people relationships are important. Sometimes, you can go on a project and have people with bad attitudes; they're all under the gun. Learning to get along with people, of course, comes with age. Some of these young people, they just haven't reached that level yet...To get more out of your people, your employees, be nice to them.

Following His Father's Example

I've learned it the old way from my dad, because I would see how he handled his books, how he handled his finances. My dad never charged anything. He would pay everything at the end of the month. He had all his bills paid, and no credit cards. He didn't charge things or overspend.

One day I told him, 'Dad, we need a truck,' and he asked, 'How much is it?' I would tell him and he would say, 'Okay.' When he had his money saved up, he would buy me a truck. That's the way he did business. His equipment was the same way; we started buying better welding machines, better tools. I'd say, 'Dad we need this.' He'd say, 'How much is it?' I'd say, 'Dad, it's so much!' But I knew what he was going to do; he was going to save up for it. He would say, 'If you have to buy on credit, you can't afford it.' But it's a different era now. These days you have to have a line of credit.

Getting Clients Out of a Predicament

I've always had a good rapport with some of the superintendents on the jobs. We get them out of the fire sometimes. One example was a hotel in San Antonio. It was the last part of the project and they needed some aluminum handrails. The superintendent forgot about the aluminum handrails and it's close to the time when they have to get a Certificate of Occupancy. So who do they call? We quickly got them their handrails; we got them out of a fire.

Workload and Workmen

Othon does not advertise; he says he doesn't have to.

In the past few months, I have turned work away; I just tell them, 'We have projects here that we've started.' And I told my workers, 'No, we're not going to take on anything new; we're just going to finish what we have.' I can't be under the gun day and night. We used to work until three o'clock in the morning. We had to finish

Othon's father, Juan Garcés, and his handiwork in 1924 at the Solino residence, San Antonio, Texas. (Courtesy Garcés Designs)

these projects by four in the morning and then come back again at seven or eight that morning. I said, 'No, we're not going to do that anymore; we need the rest.'

I train as many as I can, whoever wants to learn. Right now we have three employees, all from San Antonio, and that is enough to carry on the workload we have right now.

What? Retired!

At the time of this interview, Othon was sixty-seven and said he was retired, but he still gets up at 5 a.m., and on this day, he had already been working at his shop for over ten hours.

> *Yes, this is retired. I think I work harder now that I'm retired than when I was working, but the day I fully retire, I think I'm going to be sick. That's when I'm going to retire: when I'm six feet under! But I hope that isn't soon.*
>
> *Actually, when I* fully *retire, I would like to do artwork—abstract art—in metal, sculpture and things. I have ideas. My work demands so much of me that I really don't have much time. I've had friends who were welders, just plain welders, and they started doing artwork and they've done much better financially. Just doing commissioned artwork, you know, doing special projects.*

Othon Garcés has been working in his craft for more than sixty years.

> *I always loved it, you know. I wouldn't be in this trade if I didn't.*

"Learning to get along with people, of course, comes with age. Some of these young people, they just haven't reached that level yet . . . to get more out of your people, your employees, be nice to them." —Othon Garcés

The Gate

Othon tells this tale with a smile of warm pleasure:

There's a gate that used to be on the grounds of the Solino residence. A number of years ago we were called to remove the gate and put it in another location. I told the architect and the landscape architect that my dad had made that gate. And they kind of said, 'No, your dad didn't make this gate!' So, at the next construction meeting, I take out this photo, and it had a date that my dad put on there—1924. And this was sixty years later, in '84. I still remember. I have a snapshot; I took a picture of myself in the same location where my dad was. That gate was in perfect shape. All we had to do to it was sandblast it. We repainted it and reinstalled it and that gate was like brand new. No rust on it, not a bit, and everything was intact, nothing we had to do. Just take it off and put it back up and paint it.

The same gate in 1984. The residence is now the Luther-Brown Estate. The wrought-iron jewel has been moved to the rear of the property. (Courtesy Garcés)

Those days you had, I think, better craftsmen; we're too much in a hurry now. Not because we want to be; they make us hurry. Sometimes you just don't have the needed time to do some of these projects.

Cost, Sources, and Quality of Materials

The cost of doing business nowadays, well, right now, we can get a quote for materials and sometimes it's only good until 5 p.m. or until tomorrow. Materials costs are actually going up that fast! And supply and demand—there's more demand for it now, so the price goes up. The materials we use come from all over the world.

This is one of the things that I always point out to my employees—all the materials that come from overseas have a stamp on them or a tag. They come from Taiwan, from Canada, from Turkey, from China. A lot of the castings that used to be manufactured in the U.S. are now coming from China. If we do government work—which we do very little of—the material has to be certified. If it has to be U.S.-made material, they specify that. Plus, you get your certification. For most of these projects, though, there are not high demands or very strict regulations. In some cases, the contractor just goes with the cheapest price, which most of the time doesn't work. On the other had, some of these factories overseas have really improved. You get a really good price. You also know they have imperfections, but you know throughout the years they have been able to refine a lot of their products.

Melinda Garcés, President of Garcés Designs, and her father, Othon, at one of their newly fabricated "Urban Porch" bus shelters [Collection of the City of San Antonio, commissioned through "Public Art San Antonio"]. (Don Kirk)

His Favorite Materials

Othon prefers to work with wrought iron and steel.

The advantage of wrought iron is that you can be more creative. It lends itself to forming a lot of these shapes. With materials like brass, you cannot forge it, you cannot bend it the same way. Wrought iron has a totally different feel, a different effect. I think in wrought iron you put more of your heart into it because you work the material. You work it to make it be what you want it to be. You forge with a hammer. The material comes to you as sheets or rods. You put it together by welding it. You can even twist it. The sheets come in different forms—round, squares with raw angles and so forth. You just form it into the product that you want to make. It can be a table, it can be a window grill, it can be a light fixture.

Other Important Public Buildings

Othon has worked on either original creations—or restorations—for many famous San Antonio landmarks.

We did a lot of the work at Mi Tierra. All of the wrought iron. The Mariachi Bar, that big arch, that wrought-iron gate . . . we did all those for the Cortez family.

At the time of the interview he was working on a project at Mission Concepción with the architectural firm of Ford, Powell & Carson.

We're doing some lighting supports with Karen Peterson.

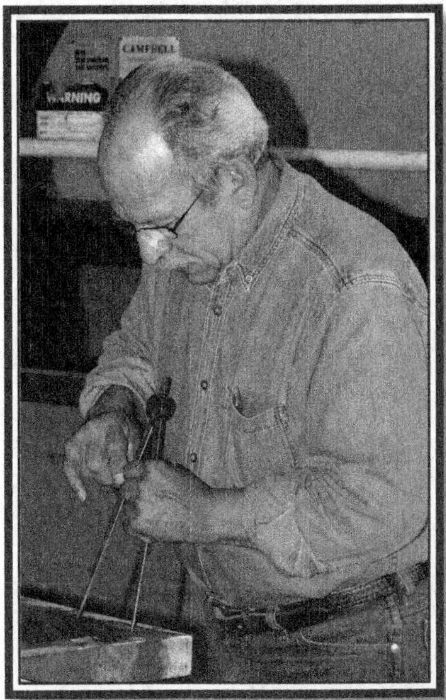

Othon using dividers to equally space design elements. (Monte Adams)

Othon also worked on light fixtures for the McNay Art Museum, as well as working with Grant Martin on the Witte Museum Treehouse. He produced special chandeliers for the Southwest School of Art.

We made them from scratch, from a photograph. Ruth Johnson came up with that design. That was in the '80s. There are also light fixtures for Club Giraud and for the Maloney's.

He describes a job for United Services Automobile Association (USAA):

The people who were doing the cushions got frustrated and said that nobody here in San Antonio could make these legs for a bench in the colonnade at USAA headquarters. They went to a supply house, and they recommended us. They came up with the design, we came up with the technique on how to put it together.

Othon says his work and satisfied clients are his best promotion.

She'll say, 'I want you to get Garcés to do this work,' so that's my advertisement; that's all I need to stay busy.

MI TIERRA CAFE & BAKERY

Othon's wrought-iron creations at Mi Tierra in Market Square, San Antonio, Texas. (Don Kirk Photos)

MI TIERRA CAFE & BAKERY

The San Fernando Cathedral Has Special Memories

The wrought-iron cross atop each bell tower of San Fernando Cathedral was constructed by Othon and his dad. (Don Kirk)

The two crosses that are up there at the very top on the outside, my dad and I made those. I helped him forge them. When we were working at the cathedral, my dad would send me over there because it was on a Saturday. He'd say, 'I want you to go over there and get the guys going and start setting up the fence.' The team boss had his camera with him all the time, and he went in there and asked, 'How are you doing?' So, we started looking over the job. We walked around and came outside and he said, 'Othon, where is it?' And I was wondering if he had lost something. 'Where is what?' I asked. 'What are you looking for?' He replied, 'I'm looking for a mistake! Where is the mistake?' And I told him, 'You're not going to find a mistake because not only will I get it from you if there <u>is</u> a mistake, I will also get it from my dad!'

That was when we renovated the cathedral, the first big renovation. That was in the '60s—'68 or '69—something like that. They took down all of that wrought-iron fence, and then they put the stone fence around the cathedral and the little rectory house. We reworked all of the ironwork, straightened it up, cleaned it up, repainted it.

It was old metal and interesting to work with because you could see all the rust. But once you cut it, it's like brand new. It doesn't corrode. Unless it's underwater or around a lot of salt or something, but other than that it lasts. Also, before they took out the chandeliers, there used to be some brass chandeliers in there, and we made those. Now they're at the new reception center they have right next to the cathedral.

Passing On the Knowledge

Othon would like to pass on his skill with tutoring, and he is saving some tangibles like drawings and files:

> *I do have some old drawings that I worked on. As a matter of fact, I still have a ledger of my dad's. He would write down who he did the work for, how much he charged, and when they paid. I want to keep that with me for a long time and pass it on to the kids. I've got a little hammer that was my dad's. I've got one of his rulers. It's an old style, all in inches, made by the Lufkin Rule Company from Michigan. I'll save them.*

Othon's hopeful that once he's truly retired, he can pass some of the papers, and perhaps the tools, on to

> *that new Smithsonian affiliate, the Museo Alameda. I'm sure someday, sometime, they're going to ask for some of the work of craftsmen in San Antonio, because we work with the Cortez family. He has mentioned that someday we're going to have some of the artistry from San Antonio displayed at the Alameda, at the Smithsonian. We're the ones who have helped create San Antonio, keep it together. I don't take a lot of credit. Just to say <u>we</u> did that! That makes me happy.*

The always-jocular Othon Garcés. (Don Kirk)

Othon does not believe the need for his type of artistry will ever cease.

> *There's plenty of work; that's for sure.*

The question for us is: Will we have artisans of his stature in the future to whom we can safely entrust the preservation and restoration of the most beautiful and sacred monuments of our shared heritage?

'Urban Porch' bus shelters built by Garcés Designs in 2014. A Collection of the City of San Antonio, commissioned through 'Public Art San Antonio'). (Don Kirk Photos)

Othon supervising final touchup as Angel Gonzalez (on the left) and Reginaldo Cruz do the work.

Othon Garcés posing for a portrait at one of the new bus shelters.

URBAN PORCH

"It has long hours and you might say the pay is not the best, but when you see something completed, that's the reward—that is your pay." —Othon Garcés

URBAN PORCH

This bus shelter actually has a swing! (Don Kirk Photos)

6. CURTIS HUNT & FAMILY

(Tommy Adams)

INTRODUCTION

CURTIS ROLF HUNT III is the go-to guy for the masonry restoration of historical buildings in San Antonio, South Central Texas, and beyond. Master masonry has been the family passion for generations. "Right now, my sons are in it; I'm in it; my father was in it; my grandfather and my great-grandfather were in it."

He and his wife Dorothy, who first met in the eighth grade, have been married over thirty-five years. Each of their three children is an important contributor to Curtis Hunt Restorations, established near Elmendorf in 1978. His sons, Michael and Curtis Hunt IV, are project managers, and daughter Brandy helps Dotty run the office.

One of dozens of important projects is the restoration of the Pearl Brewery, dating from the late 1800s. For Hunt and his family, the future is clear. "There are so many old buildings that there's always work to be done, especially in San Antonio, with its history and heritage."

As you read these pages, you'll see that the Hunt family has done much to help preserve our history and heritage for generations to enjoy.

(Tommy Adams)

Curtis Hunt.

Curtis Hunt & Family

MASTER STONE MASONS

Curtis Rolf Hunt III, master stone mason, embodies both his family's lineage and that of his ancient craft. We interviewed him under a hot August sun at a recent work site, the restoration of the Pearl Brewery. Our "backdrop" was the boiler room, built in the 1890s.

A San Antonio native (born at Santa Rosa Hospital, January 15, 1952), Curtis has never been far from home. He attended East Central High School and met Dorothy, his wife-to-be, in the eighth grade. They started dating in the tenth grade and married at age nineteen. Married over forty-four years, Curtis and Dotty have three children—and each one is an important part of the Hunt family masonry business.

Hunt's wife has been in the office since he started on his own in 1978. His daughter Brandy works in the office too, and his two sons are project managers. The office/home is located in the far southeast part of San Antonio, near Elmendorf, Texas. The Hunts live on the same property as their business, Curtis Hunt Restorations, and are really never "off duty."

> *That's true, we get calls all night long and start in about 5:30 in the morning.*

Curtis grew up in the family business.

> *Right now, my sons* [Michael and Curtis Hunt IV] *are in the business with me.* [As a young man] *I labored during the summer for my dad and my uncle.*

Curtis' father passed away in 1983; his uncle died in 2014.

Interesting occupations run on both sides of the family. Dotty's father, retired now, was a Borden's milkman delivering to customers in the Pleasanton area for many years. Curtis' brother worked for the National Park Service on such important historical buildings as Lincoln's log cabin. One recently completed job was restoring Fort Jefferson, built in the 1850s on one of the last islands off of Florida. Working for the National Park Service, he was the project manager for the two-year effort. Based in

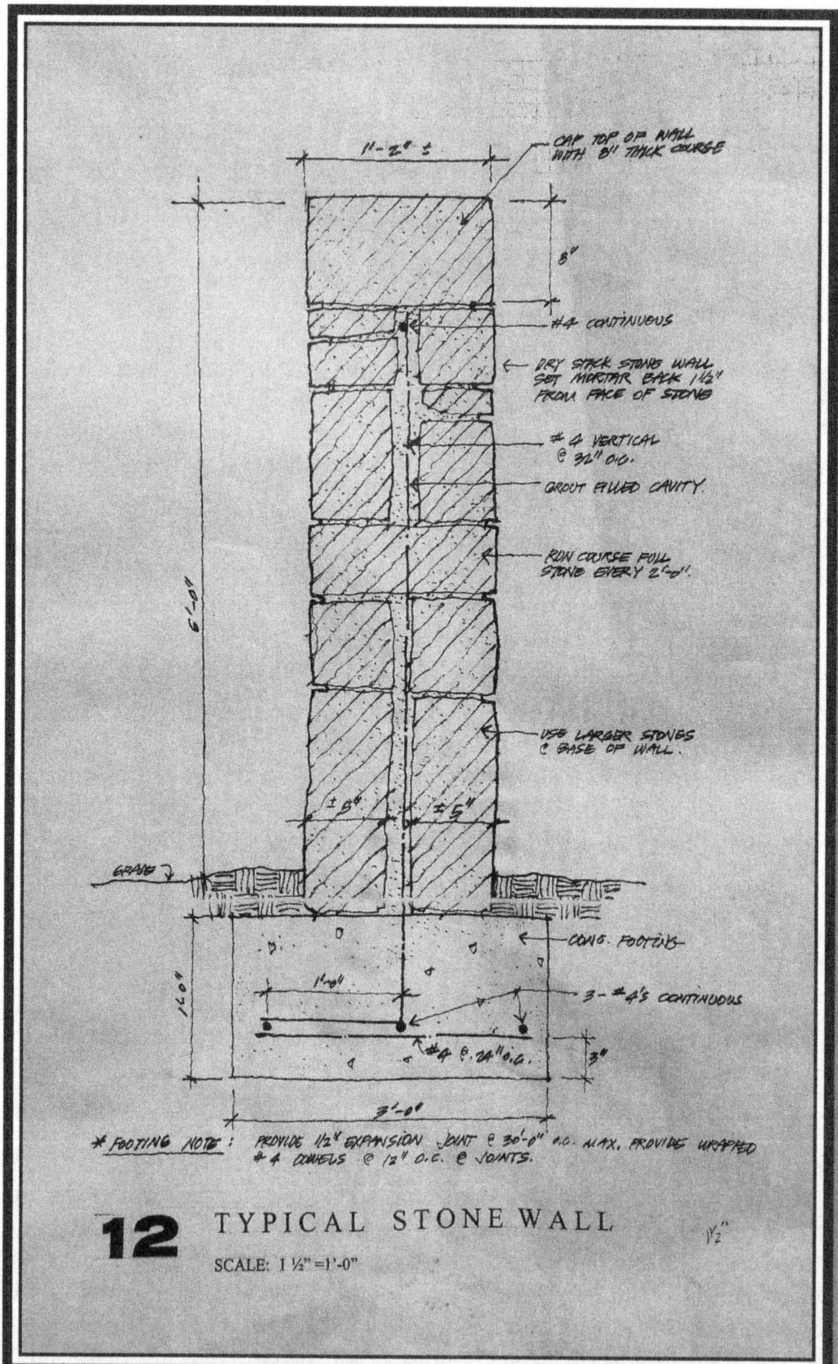

(Courtesy Rubiola Construction)

Atlanta now, he returned to the family craft after a stint in the Navy and some Civil Service years in Corpus Christi. Like Curtis, he was trained by his father and uncle as a master stone mason.

> *There are so many old buildings that there's always work to be done, especially in San Antonio, with its history and heritage. In addition, you get all the little towns that are around—Fredericksburg, Kerrville—and farmhouses out that way. Lots of restoration to be done.*

A Family Business

As a Hunt, Curtis is the continuation of a family of masons dating back generations, including not only his father and uncle, but his grandfather and great-grandfather.

> *Actually, it goes a little farther back than that. We came from Pennsylvania down into Kentucky, and then my great-grandfather came from Kentucky to San Antonio and landed here at a brick-yard. I think he went over to San Diego, California, for a while and established a brickyard; then he came back. My grandfather was born here in San Antonio. My great-grandfather built several buildings around here and, of course, my grandfather went into the business. My grandfather and I were real close.*

The Historical Lineage

As stone masons, the Hunts practice one of the oldest professions on earth. Curtis has pride in the heritage of his craft, which dates back thousands of years—to when people began building permanent edifices.

> *I think our trade came well before the architects—they claim the number two profession, right?*
>
> *Even before they built the pyramids in Egypt, or the temples of the Mayans, or Aztecs and Incas in the New World, masons were busy creating the first adobe homes, stone steps, and masonry meeting halls.*

As a masonry contractor in today's world, Curtis Hunt helps preserve the legacy of previous master masons by doing his part to restore historical buildings. At the time of this interview, he was well into a two- to three-year project *"restoring all these old, beautiful masonry buildings"* at the former Pearl Brewery in San Antonio.

Two of Curtis Hunt's men restoring a stone entrance column on Crockett Street in downtown San Antonio. (Tommy Adams)

Getting Started

Curtis started learning the masonry business while still a boy.

I started at about fifteen I think. And even before that, my brother and I unloaded trailers when we were around ten years old. Dad would come in with a trailer-load of scaffold board and scaffolding and we would go out there and unload it. He would give us a couple of bucks. And then I can remember building for Tom McCanns, laboring on Pleasanton and Military Drive. I'd drive over there at fifteen. I'd take some friends of mine and we would labor.

Although he helped his grandfather and father with masonry jobs as a youngster, masonry was not Curtis' first choice for a career. His first job when he finished high school was working in the oil fields.

> *I actually got burned up; I hit an electrical line and got put in the hospital. And then my father came and said, 'No more of that; get out there where you belong* [in the family masonry business].*' Plus, my aunt jumped me. She said, 'What am I messing with holes in the ground for when I got all the walls I can build in the world?'*

A Family Tradition

His grandfather built masonry buildings throughout San Antonio including one in St. Paul Square and a few at the former Pearl Brewery site where Curtis has been working recently.

> *My grandfather told me the story about when he was working by the day. He got a job—contracted to do a chimney for like four bucks. That was in 1908 or 1910; he built it in two days making*

only a dollar a day. Good money back then. Then he got married and had my dad. My dad was born in 1916; he grew on up, got into business with his dad and my uncle.

As far as working with his father, Curtis remembers that, for safety's sake, he didn't get on the wall after the mid-1970s.

I got to lay brick with him one time in the 1960s.

Curtis' grandfather retired at sixty-five, but he lived to be ninety-two.

I got to lay brick with him too.

Heritage Buildings

Almost every iconic "heritage" building in San Antonio reveals the imprint of a Hunt in some way. One of Curtis' greatest thrills is the fact that he often finds himself working on the very building one of his forebears first built.

Curtis helped restore City Hall in '93 and '94.

What's really interesting about City Hall, my grandfather put in the fourth floor in '27. I got to repoint the mortar joints he had built.

As for the Alamo, Curtis' father worked on it, as did Curtis.

Supposedly, my grandfather did a little work when he was sixty. I worked for my father at the Alamo in about '74. Then in '93 or '94, I worked on the front and side and did some other things around there—maintenance on the building itself. Pretty neat.

His brother worked on the Royalty Coins Building "way back." His uncle helped with the restoration of the Southwestern Bell Building.

They totally restored all that terracotta.

Curtis and his family also worked on the Sullivan Carriage House, helping dismantle it, move it to the Botanical Garden, and rebuild it completely.

Learning the Craft

Hunt stresses that though there are some classes at the University of Texas and other schools, learning this craft basically takes on-the-job training.

There's nothing better than just getting out here and getting with somebody and starting to do it. You start out with what's considered a "cub" or a two- to three-year apprenticeship and then after three years of brick, rock, whatever, you turn out as a mason or bricklayer. You still don't achieve the skill of a true mason for another two to five years after that. It takes at least five or six years to really understand masonry work.

The 1893 Clifford Building in downtown San Antonio (aka the Royalty Coins Building) was worked on by Curtis' brother. The Romanesque Revival-style building was originally known as the "Riverside Building" and was designed by architect J. Riely Gordon. (Don Kirk)

In the Hunt family, of course, the craft is passed from generation to generation. Curtis also had some other teachers, including his Uncle Bubba and others.

> *A real nice guy named Bill Rosler taught me speed as far as new construction. There was an older fellow named Reymundo Rodriguez ,who had been in the company longer than I was. He came in at about '62 and he taught me a whole lot about taking care and laying stone and what to watch for. He taught me how to have patience when you do restoration. It took quite a while before I could figure out what Reymundo was telling me.*
>
> *We have what's called the San Antonio Masonry Contractors Association (SAMCA). We have an apprentice program that we've set up, but we also have gone out to Southwest High School and built up some brick walls on Career Days. We had a seminar and showed them how it's done, but this trade is more about coming to work as a laborer, then working your way up from that point.*

Learning About Restoration

> *A lot of the restoration that I've learned is actually from two lady architects, Carolyn Peterson and Annie Seltzer. I've learned a whole lot from them because they're real good. They really know restoration. It was real nice when you could ask them how or why, and they would explain to you why we were doing it this way or that, why we're doing chemical injection. You could get a lot of technical information from them.*
>
> *When you're a craftsman, you do want to know why, don't you? That's what I try to teach my sons: Don't hesitate to ask why, because then you'll gain the knowledge of that person.*

There are not a lot of people in San Antonio doing what the Hunts do. Curtis says he knows perhaps ten in Texas.

> *Most don't have the patience for it. Most major contractors want to get in and get out and go on to the next one. But restoration is just not that way. I actually had a friend run out of work and he sent a couple of his masons over here, and they lasted one day. They said it's just too slow. I told them, 'You know I'm paying you the same,' but they just said they didn't have the patience for it—patience.*

In order to find masons who can be successful at restoration work, the Hunts basically have to train them because it requires both special expertise and time-consuming attention to detail.

> *And, it takes a certain type of person.*

(Don Kirk)

THE SOUTHWESTERN BELL BUILDING

Curtis Hunt's uncle worked on this building, restoring a massive amount of terracotta so that it could be enjoyed today.

One of the Hunts' long-time employees went to school with Curtis and came to work with the Hunts in 1978 after his service in the Navy. He, like other Hunt masons, finds the work rewarding and appreciates working for, and with, such dedicated craftsmen.

The Challenge of Restoration

> *The difficulty that you run into is that there are so many unknowns. You get into this building and start restoring it and you may run into some structural things that you can't tell until you get into it. Sometimes you have to literally dismantle walls and take out what they put in—like pipes and I-beams—such as that one up there that we have to take out. Put it back to its original opening 'cause through the years they altered the building so much, they changed it. The goal is always to bring it back to as close to original as possible, exactly how it was when it was built. Our goal is restoration.*
>
> *There are certain rules and regulations that you have to follow to stay within restoration. For instance, our mortar has to be a high lime content mortar. You can't just come in and pour cement because it will damage the building. The mortar can't be any harder than the material that you're pointing up—for instance, brick or anything like that—because it gets brittle and has a tendency to bust the brick. The Texas Historical Commission has guidelines we try to follow.*

The Building Blocks

> *Masons build buildings out of brick, stone, block, CMU—basically anything out of masonry.*

CMU stands for "concrete masonry unit," the gray cinder or concrete block that you see used in structural or "curtain" walls on many commercial buildings. Curtis and his employees work with a lot of clay tile, too.

Curtis says just finding the materials can be hard sometimes. There can also be difficulties trying to match the original material, no matter what type of masonry. For example, Curtis searched far and wide to find matching brick for the Pearl Brewery restoration.

> *I finally found some in Corpus Christi on a building that was taken down there, so we purchased 15,000 bricks to come in here and help take care of this, and I'm looking at the building now that the paint's off, and we're going to have to get still more brick.*

The Tools

Working on historical restorations requires more than patience; proper tools are also critical. As Curtis is quick to point out, the craftsman is not allowed by rules and regulations to use mechanical devices to dig out the mortar joints or to use electrical tools of any kind.

> *You can't just come in and start beating a wall to pieces. You have to be real careful in what you do. If a guy is in here with a grinder, he may lose it and he can tear up the face of the brick or whatever material you're working on. You have to go in with a hammer and a chisel and gently remove everything, just so. We cleaned the stone at the Alamo—the cornice work—with a toothbrush!*

There are some very modern tools used today in this ancient craft. The Hunts buy their crack-repair adhesives from Edison Coatings or Cathedral Stone Products.

> *They'll send us the package deal with the needles and things that are required to use with it.*

To learn to use the materials correctly may require special training.

> *We actually flew a guy down who trained and certified five of my guys to do stone patching.*

Juan Torres cutting a CMU (concrete masonry unit) with a masonry block saw, Southwest School of Art, San Antonio, 2014. (Don Kirk)

As far as handed-down family tools, Curtis has one tool his grandfather gave him for setting stone.

> *It keeps your fingers from getting crushed—a pretty cool little tool. If you lose your fingers, you're out of work.*

Restoration of brick requires removing layers of paint and replacing unsound grout. (Don Kirk)

Cutting-Edge Technology

Curtis Hunt and his company artisans are working with the very latest techniques.

> *We're getting into consolidating stones and bricks. It's a special chemical we're injecting over at the Pearl Stables. It's very scientific, actually. We injected a damp portion of fluid to stop the moisture from coming out of the ground. You have to do it a certain way. You have to drill holes every six inches and let gravity fill it.*
>
> *We're doing a lot more with crack injections. We buy some materials out of New York and out of the Carolinas. We actually use hypodermic needles to inject fluid into these cracks and fill the void up so that the moisture can't get in there and cause what's called a 'Freeze-Thaw.' That's when moisture gets into it and if it freezes, it expands and can crack your building even worse. It's becoming so much more technical than it was twenty years ago.*

As always, the Hunts work with the old and embrace the new.

> *It's kind of interesting, but my sons are really involved in the new technologies, and that makes it nice because I'm from the old school that just gets in there and chisels it out. Like I say, with the new architects, new techniques are being developed and we have to keep up with them.*

The Pearl long before the current restoration. (Don Kirk, 1980)

Imagination, Logic, and Developing an Eye

Regarding achieving mastery, Curtis explains:

You have to have an eye for it because the picture is a puzzle. You're putting up this wall and you have to look at a stone—really look at it—and see if it's going to fit into the area you want it.

There are some guys out there like Reymundo Rodriguez. I'd be laying stone and I'd be beating my brains out trying to shape this stone and he would just look over and say, 'Get that one,' and it would fit with very little work. So, you have to

Curtis and his son "Cooter" (Curtis IV) at the Pearl Brewery restoration discussing the next task: where they're going for lunch. (Barbara Dean Hendricks)

The tower of the Pearl Brew House, camouflaged by scaffolding. Curtis found himself restoring his own grandfather's work. (Tommy Adams)

train your eye to see what your next move is. You've got to get your blinders off if you're going to do this type of work, especially.

And, you've got to use logic. Sometimes they get out there and it doesn't make sense to me, and I have to say, 'Wait, hold on just a minute, why are we doing it this way?' My dad taught me to look at different views—don't just look at it one way. My cousin Shard was telling me a story the other day about how he walked by and my dad (who was called Boogie) said, 'Shard, come here.' And Shard said, 'What is it, Uncle Boogie?' He said, 'What do you see?' And he held a bicycle chain up and Chard said, 'Well,

"So, you have to train your eye to see what your next move is. You've got to get your blinders off if you're going to do this type of work." —Curtis Hunt

THE RESTORED PEARL BREW HOUSE TOWER

THE PEARL IS NOW A "CULINARY AND CULTURAL DESTINATION" IN SAN ANTONIO.

The Pearl Brewery was founded in 1883 and closed in 2001. Purchased by Silver Ventures, Inc., it has become a focus of revitalization efforts in northern downtown San Antonio. (Don Kirk)

Working high on scaffolding, laborers carefully removed many layers of paint from this historic structure. Every square inch of the original brick walls had to be stripped of many layers of paint and the mortar joints repointed as needed, 2014. (Don Kirk)

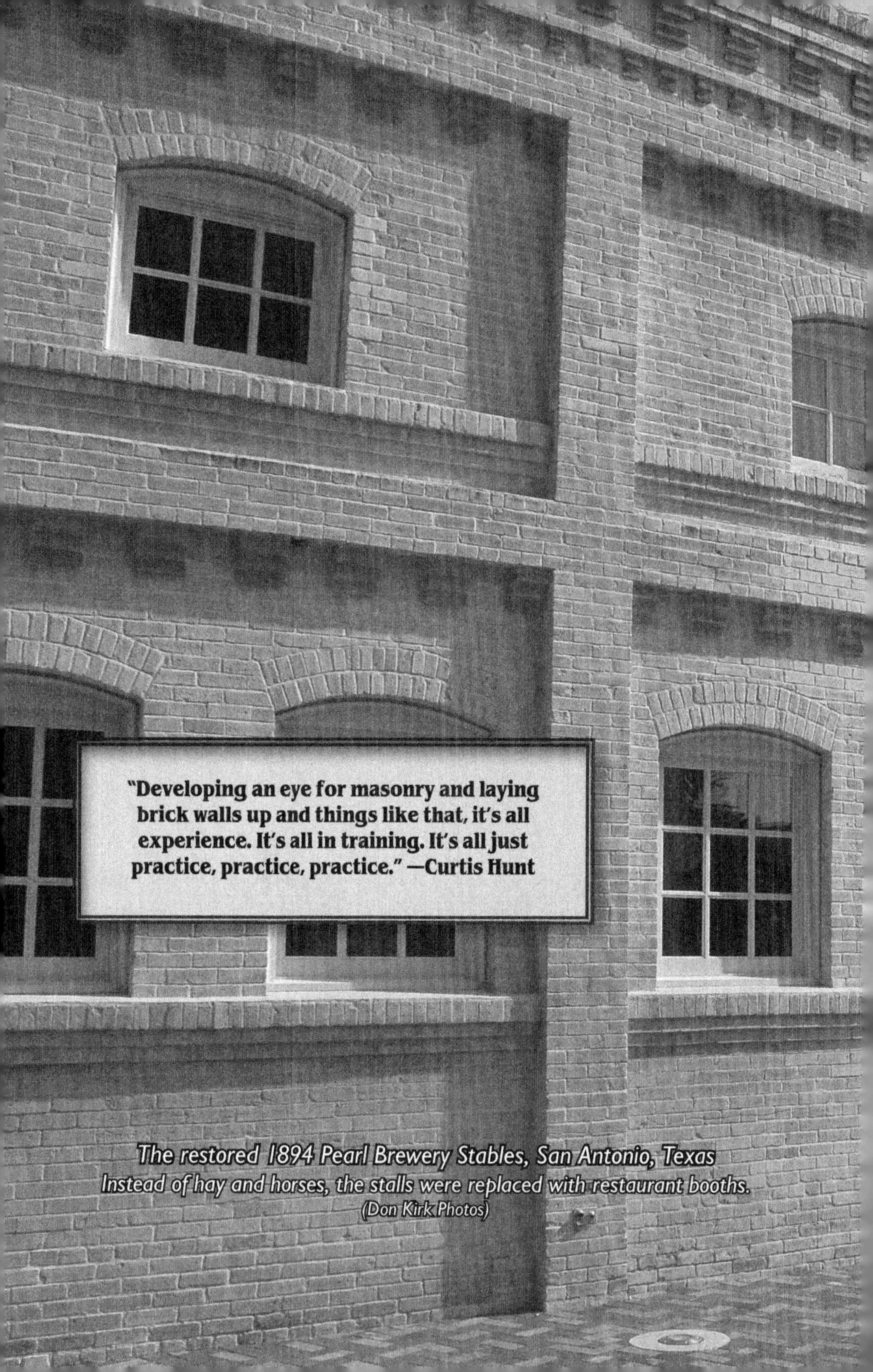

"Developing an eye for masonry and laying brick walls up and things like that, it's all experience. It's all in training. It's all just practice, practice, practice." —Curtis Hunt

The restored 1894 Pearl Brewery Stables, San Antonio, Texas
Instead of hay and horses, the stalls were replaced with restaurant booths.
(Don Kirk Photos)

1894
STABLE

Reading "acres" of blueprints required for even the simplest of jobs. (Don Kirk)

it's a bicycle chain.' He said, 'No, see beyond that.' He said, 'With this thing we can lift brick.' And he designed one. He bought a big, long chain and put a drill with a sprocket on it and put hooks on it and he used that to lift brick up to the scaffold. 'See,' he said, 'you got to look beyond what you think it really is.' That's what my dad taught me—how to lift things with ease with hoists and rails and things like that and that's why I really like doing that kind of thing.

Grandpa would always say 'lay in the valley' and things of that nature because you got a point and you can't lay the stone on top of that point. You have to lay in the valley to make that wall go on up. Otherwise, it will just fall over on you. The critical part of being a stone mason is training your eye to see what type of stone fits and working that into the wall.

Yes, getting trained in drawing, drafting, and algebra can help when you are trying to lay out a building and trying to get your different heights. And your windows have to be at a certain height; you have to course out your brickwork to work to hit that window height so you don't end up with a cut. But, developing an eye for masonry and laying brick walls up and things like that, it's all experience. It's all just practice, practice, practice.

In addition to practice, part of the mastery of any craft is learning the language. In any craft, there are technical terms for processes and techniques.

On the Wall

'On the wall' means, obviously, you get up on a scaffold and start laying brick. It's just slang for laying brick, getting up 'on the wall' with the guys and laying brick all day long.

Laying brick with my grandfather, I was building a chimney at one of his rent houses and you know—naturally, coming from a family of bricklayers—you have to do it right or you get chewed out, so I was taking my time and making sure each course was done. I was making a flue and my grandfather said, 'You'll never make it, son, you're too slow.' I told my dad, 'You know, if you both go away I can get this thing done,' but I'm worried about him chewing me out 'cause I think I made a brick out of level.

Buttering a Brick

Curtis explains this technique with another story.

A friend of mine, Travis Crow, his dad was a masonry contractor who worked for my dad way back when. But anyhow, when his dad came out, he made him take two saw horses and a two-by-4 and all he did all day long was spread mud on the 2 by 4, just spread mortar. Then, he'd turn it edgeways and then spread mortar on a block. He did that for two or three days just to get the hang of how to spread mortar 'cause that's 50 percent of new work—spreading mortar on the wall. That kind of training is what you've got to have, not as far as any books or anything like that. It's like driving. You can read all you want about driving a car, but until you get in that car, then it happens.

Curtis "Cooter" Hunt supervising an addition at Southwest School of Art. (Don Kirk)

Coursing Your Masonry Out

This is coursing [he points to an opening above]. *You're 'coursing your masonry out.' In other words, you're going to go across that opening up there. So what you want to do is take a ruler and get the top of the window and measure it down, so that—if you have to—you can adjust these mortar joints so that when you get to the top of that opening, you don't have to put a cut. You just go right on across. But you have to figure that out ahead of time, especially if you have a bunch of windows or old windows.*

A Roman arch over a small window, relaid stone-by-stone at the Sullivan Carriage House in the San Antonio Botanical Garden. And below, a Roman arch over a door, DeWitt County Courthouse. (Don Kirk)

The Arch and the Keystone

Some arches have a big stone in the center called the 'keystone.' But, with every arch, you start in the center and then you work yourself left and right from that center brick or stone, and that's 'the key' and it locks that arch in place. Then, if you have to adjust your mortar joints when you make that arch, there's always one brick dead in the center of the opening.

The critical keystone keeps the arch from collapsing, allowing the load above—maybe a wall—to span the opening. The load is carried to each side—sometimes on columns—and down to the ground.

Let's say this is a two-story building and you have a set of windows up there. You have to open your coursing up or you may have to tighten it up to get to this first window. Now you have to scale for the next window. You may have to do the same up or down, but you don't want your mortar joints to be more than an eighth of an inch because otherwise they look too big. You try to maintain a 3/8″ mortar joint. That's the kind of stuff that you learn out in the field.

Restoring an Arch

There is nothing prettier than an arch. Masonry art is just beautiful when you look at it. You can do anything you want out of stucco, glass, or whatever, but you can't take the place of the character of an arch.

To restore an arch, we dig out all the mortar joints where they patched it through the years. For example, a seam patch—we'll take that patch out and actually dig back four inches and put the new brick in there—not brand new brick, but brick to match—and put it back to its original.

"There is nothing prettier than an arch. Masonry art is just beautiful when you look at it. You can do anything you want out of stucco, glass, or whatever, but you can't take the place of the character of an arch." —Curtis Hunt

Standard Sizes

The standard thickness is 3/8″. The new work is all a modular system. Everything works to eight inches, you can take eight inch cubes and that's considered a modular system. Brick, CMU, clay tile, all that's designed to be in eight-inch increments.

Unique Challenges

Every project is totally different. You come in and you start evaluating what has to be done. There are so many unknowns in restoration. You never know what you're going to get until you get into it. You may open up this wall and find cracks and 'Oh, wait a minute,' you have to stop and talk to an engineer and say 'Look, this thing is fixing to fall,' so you have to go in and reinforce it.

A lot of times you'll talk to an engineer and say, 'Okay, this is how we've done this before. In this type of application, we drilled stainless steel pins through here and locked that crack back together in this way.' It just depends on each project. It's just so many unknowns that it's unreal. A lot of people don't understand that.

Augustin Perez "on the wall" troweling off excess mortar on freshly laid CMUs. Below, Juan Torres with the heavy work: mixing up a large batch of mortar, Southwest School of Art. (Don Kirk)

A Favorite Project: The Sullivan Carriage House

Curtis is very happy to have worked on so many historically significant buildings.

Dismantling the Sullivan Carriage House was a real nice project, a real neat one to do. We numbered each stone as we dismantled it and documented it on photographs and working drawings. We documented the dimensions on the walls themselves—the lengths and everything else. We got that thing back in within a half-inch, I believe, with all those arches. That was pretty cool.

A Favorite Project: Wharton County Courthouse

Another project that I'm real proud of is the Wharton County Courthouse in Wharton, Texas. If you could see what it looked like before we started and then the end product, it's just amazing what we did there.

It was originally a beautiful building—with stone arches, stone entrances, and all that. In 1935, they came and knocked all the stone off that was projecting out of the wall except the corners. They kept the quoins and plastered this whole building in art deco.

BEFORE: A view of the Sullivan Carriage House in 1980 on San Antonio Light property. In 1988, the Hunt's dismantled and moved it, stone-by-stone, to the Botanical Garden, where it was completely restored and opened as a restaurant and gift shop in 1995. This beautiful Richardsonian Romanesque-style coach house and stables was designed by architect Alfred Giles in 1896. (Don Kirk)

In the 1930's and 1940's, the courthouse was remodeled to gain needed space; the bell tower was removed and stucco was used over the original brick. Additions to the first floor of the building resulted in a white stack of ugly rectangles.

> *We went in there and carefully removed all the plaster and put all the stone back. We actually had to cut the plaster in grids with a saw, but you could only set your saw to a certain depth, because you couldn't cut into the brick. They let us cut slit grooves in the plaster. Then we came in and tapped it real carefully, and peeled it gently off the face of the brick.*

Like many Hunt projects, this was not a short, simple job. Phase One, the exterior, took about three years, and—at the time of this interview—craftsmen were restoring the historic interior.

Working from detailed plans, craftsmen carefully removed, numbered, and organized each stone of the Sullivan Carriage House for transport to the new site at the San Antonio Botanical Garden.

A special tool was designed by Curtis Hunt to lift the stone blocks without damaging them.

During the dismantling of the blocks, elaborate bracing was used to protect against a wall collapse. A lime and mortar mix had been used for all the mortar joints; only the foundation had been constructed with Portland cement. Salvadore Vasques is seen in the lower left-hand corner.

Photos Courtesy Doug Pape, UTSA Libraries Special Collections)

SULLIVAN CARRIAGE HOUSE

Many of the timbers from the original structure were reused. A foundation stone had to be recreated to rebuild the unique twin chimney. Once the exterior of the structure had been reconstructed, another eight years were required to complete the interior. Othon Garcés was one of those craftsmen who helped adapt the interior as a restaurant and gift shop using his wrought-iron skills. (Doug Pape, UTSA Libraries Special Collections)

AFTER: The completed structure in 1987. "When we were done, we were less than one-half inch off from the original and every stone was back in its original place." —Curtis Hunt. (W. Eugene George, UTSA Libraries Special Collections)

The Wharton County Courthouse before restoration, clad in white stucco and one-story additions, "a white stack of ugly rectangles." Below: The stucco was carefully stripped and thousands of bricks were needed to match the original brick.

Ripping off the new to restore the old.

(Photos Courtesy A.C. Shelton)

(Top Photo: A.C. Shelton)

The Wharton County Courthouse in 2007, after an amazing restoration. (Don Kirk)

The Doctor Urrutia gate, now at the San Antonio Museum of Art was taken apart and rebuilt by the Hunts. This 1920s work of art was moved from the former Dr. Aureliano Urrutia estate at Broadway and Hildebrand. The Spanish-Colonial Mexican tiles are illustrated with peacocks, frogs, and the Virgin of Guadalupe. Of note, the doctor, a Mexican political exile, was responsible for bringing Carlos Cortés' great-uncle, Dionísio Rodriguez, to San Antonio from México City to work on his estate. (Don Kirk)

A Favorite Project: The Doctor Urrutia Mural

Another favorite move and restoration was the Urrutia Mural—the beautiful tile and stone mural that long stood at the corner of Broadway and Hildebrand. It was part of the original Urrutia estate.

> *We picked it up and moved it to Jones Avenue. It weighed 28,000 pounds. We sawed the legs at the bottom and cut it at the top in three different pieces and then lifted the whole massive piece out. Then, we put the legs back, and put the stone back on. It was a cool job. That's the kind of stuff we like to do.*

For Hunt, the job is never boring.

> *There's always something going on. We're always getting calls for some weird, off-the-wall stuff like that 28,000-pound stone.*

A Favorite Project: The Steves' Canales

Curtis did some specialized work for Mrs. Marshall Steves.

> *I put some canales in for them. Canales are scuppers that stick out of the wall, but they're out of stone. They had broken off, deteriorated; we replaced some of those.*

"Scuppers" are devices to channel rainwater from a flat roof through a wall. They can also serve as decorative architectural elements and thus some have fanciful shapes—like cats, birds, animals—and can be chiseled out of stone. The simplest canale is just a piece of pipe, but who would want that?

Favorite Project: The Charles Butt Home

The Hunts' work is much in demand, and there's often a family connection.

> *When my father passed away, he was right in the middle of restoring Charles' home in King William. I went down, took that over and finished that up. My dad had done basically the exterior of the house; we put in the walls and things of that nature.*

Like so many Hunt clients whose contacts and projects can span generations, Charles Butt hired Curtis for a number of other projects.

> *Then he had me go up and point up the house on his ranch out by Camp Verde. The latest thing we've done is working on his lighthouse down in Port Aransas.*

Lydia Ann Channel Lighthouse, Port Aransas, Texas.

(Courtesy Charles Butt)

A Favorite Project: The Lighthouse Owned by Charles Butt

One of Curtis' most memorable jobs was helping restore both the interior and the exterior of Charles Butt's all-brick historical lighthouse on the southern Texas coast near Port Aransas, a fishing village and tourist destination.

> *That's a cool building. The pole that held the spiral stairs had broken in several different places and they patched it through the years. But the worst thing was that the steps were bolted to the wall. As they rusted, they would expand and had a tendency to crack the building up.*

Scaffolding in place. Thirty-five percent of the brick had to be replaced. A rubber grommet was used between the steps and the wall to allow for expansion and contraction. (Photos Courtesy Charles Butt)

Hunt's employees removed the pole and put a stainless steel pole in its place. They took out fifty-five steps, painted them, and had some new steps cast. Amazingly, Alamo Iron Works—the original manufacturer—still had the mold for those steps from decades past.

> *We replaced the bad steps and then we came in and bolted them to the walls, but we put a rubber grommet in between the steps and the walls so it could literally let that thing expand and contract. We took all the paint off the interior of it, too.*

The interior was finished at interview time, but Curtis still had the exterior to restore.

> *About 35% of the brick needed to be removed; all the joints would have to be dug out and re-mortared. Then there were a lot of cracks that needed to be injected, and then we needed to actually take some brick, hassle with the cracks, and just make the cracks go away basically.*

Restoration Is Detective Work

There's even a story about the original brick. The manufacturer, no longer in business, was based in New Jersey.

> *They actually lost the ship in a storm trying to bring some of the brick over. It sank like a stone and was never found.*

So the Hunts have been searching for brick to match.

> *I went over to Mississippi and looked for some brick and I found some that was fairly close.*

Curtis' brother worked on Fort Jefferson, a historic nineteenth-century fortress on an island west of Key West in Florida.

> *That brick, and the original brick in Charles' lighthouse are very, very similar. So my brother went to St. Joe's Brick and they're making some for that fort, and I'm hoping to get with them and see if I can match it up. But it's kind of difficult to find. It's kind of cool now that my kids are grown. My wife and I will jump in the car and we'll go to Mississippi looking for brick.*

The weekend prior to the interview, they went to Galveston, south of Houston, to look at a library he was going to restore.

> *Kind of gives us a little time away from everything. We do that all the time.*

A Good Reputation Goes a Long Way

> *People trust us and that makes it nice. You know that you can go do a job and they know that you're going to try to do the best you can and you have some knowledge about it. They know that our good character is there and we're really trying to do everything right. Like I said, people trust you. That means a lot.*

Like many of the craftsmen featured in this book, that trust has been built up not just by the current generation, but by fathers, grandfathers, uncles, and brothers.

You get a reputation. Dad said, 'Never, never put your name out there, never put a sign on your work. If they want you, they'll come find who you are.' That all helped out. Like I said, we got a decent reputation in this restoration business and it's worked for us. There is nothing like a family business. You know who to work with because you developed a rapport with everybody through the years. Tommy Guido of the Guido Brothers and my dad and my grandfather all worked together for years and years and years. It just keeps passing on and passing on. It's pretty nice.

Apprenticeship

Thanks to their experience and attention to detail, Curtis Hunt Restorations is flourishing, keeping everyone quite busy.

There are a lot of masonry contractors, lots of masons in this town if you think about it. Right now, probably—I'm guessing—we're close to forty, forty-five employees. We billed out twenty-one jobs last month, but I think we're down to about fifteen, sixteen jobs right now. But I'd like to get down to about twelve, so it makes life a little easier. Sometimes it seems like I'm doing nothing but running and running and running. Right now, it's at a good point and we're doing all right.

Carlos Veggara and Augustin Perez setting a corner capstone, Southwest School of Art, San Antonio, Texas. (Don Kirk)

To save the clad-in-brick 1906 Fairmount Hotel from the wrecking ball, it had to be moved five city blocks. Curtis Hunt, John Hall, and the Cavallinis all worked on some aspect of the restoration.

(Don Kirk Photos)

The hotel had to be lifted eight feet above ground with jacks and cribbing to allow hydraulic wheeled dollies to be placed underneath.

Saved from demolition by the San Antonio Conservation Society, Alamo Architects, and the Fairmount Hotel Company, the three-story, 3.2 million pound brick-and-mortar building was successfully moved and listed in the Guinness Book of World Records as the "Heaviest Building Ever Moved on Wheels."

Dump trucks loaded with sand, provide the fixed point for a block & tackle system to move the building slowly down the street toward its final destination.

The Fairmount in its final resting place at 401 South Alamo in 1986.

(Don Kirk Photos)

The seventeenth-century Mission Espada getting a twenty-first-century cleaning by the Hunts. (Don Kirk)

Since masons are not formally licensed, when the Hunts hire a new employee they generally—like many companies in the building trades—go by experience and recommendations from others. He says he might interview someone this way:

> *If they know a little bit about masonry, you can quiz them as far as what type of mortar you want to use on a particular project. What type of strength is needed in the mortar, what type of wall ties, just general information like that.*

For potential employees just considering the trade, apprenticeship, Curtis believes, is the best way to learn. But, he says, to get someone interested, you should *"try to entice him."* Hunt demonstrates how he would speak to a potential future mason:

> **"I don't want to knock a plumber, but who wants to look at a commode at the end of the day? As a masonry contractor, you can see what you've accomplished at the end of the day." —Curtis Hunt**

> *I'd say, 'Look at that wall, you get to do that.' That's pretty cool. There's not that many trades that you can do that, you know. I don't want to knock a plumber, but who wants to look at a commode at*

the end of the day? As a masonry contractor, you can see what you've accomplished at the end of the day. That's always been impressive to me.

San Antonio's Heritage

Hunt's livelihood and his passion are the historical buildings—San Antonio's three hundred years of tradition and achievement.

You drive down the road on Commerce Street and look at the old buildings. Thank God we're not Houston with all that glass. I don't want to knock Houston, but gosh, we have so much history here. Being a mason, you just envision how those guys were doing it—putting beauty into a building. The whole charm of San Antonio is amazing; I go to other cities and I don't see it. I see a little bit in Austin, but it just seems like San Antonio's got so much more character to it.

Working in San Antonio, Hunt frequently works around two-hundred-year-old stone walls and buildings.

It's pretty impressive to think about some of the stones in the walls that are four and five hundred pounds. For us to pick it up and set it in a wall we have to use derricks, but all they had were wagons and oxen and ropes and pulleys, and "come-alongs," and things like block and tackles. To go quarry it where the San Antonio Zoo is now and then haul it around and build these buildings is just amazing. You know, I'd love to go back there and just see that. I've seen pictures of City Hall when they built that and how they brought the stone blocks in from the Fair Oaks area where all that stone was quarried.

Some may think it's old and have the attitude of 'Why mess with it? Why not put up a glass building?' But you know you can't replace it. These are all solid-brick masonry walls. That type of construction is out. Today, it's just too expensive. Today, they put up wood framing behind it and then just lay a four-inch brick veneer over it. The older buildings are all solid masonry; that's why they're still here.

San Fernando Cathedral and the Alamo, both of which Hunt has worked on, are perfect examples of San Antonio's older treasures.

They built in mass. I mean, the walls are 49" thick at the Alamo, necessary to sustain the roof system and the weight and everything else. They had to have the knowledge to do all that and it was all passed down from generation to generation.

MISSION ESPADA

SAN ANTONIO, TEXAS

Michael Hunt

Ramird Aguilar

Martin Alva, Jorje Gonzalez, and Frady Armendaris mixing grout. Below: Roman Fuentes chipping out old mortar.

Martin Alva drilling holes for reinforcing.

The historic preservation of the 300-plus-year-old Mission Espada requires specific materials and techniques. Soft mortar is chipped away and replaced with grout. Wall intersections are reinforced using a specially formulated plasticized grout and stainless-steel threaded rods. Curtis' son Michael oversees the project.

Basilio "Chilo" Pizarra and Ruven Cornejo forcing in the grout injection and inserting the reinforcing rod. (Don Kirk Photos)

MISSION ESPADA, SAN ANTONIO, TEXAS

David Chaves carefully pressure-washing 250-year-old stonework at Mission Espada, San Antonio, Texas.

(Don Kirk)

Passing the Torch

> *Maybe, in fifteen years or so, I'll retire. My kids want me out probably sooner, but, you know—it's what I am.*

He acknowledges his gratitude for his work, his true passion, saying he has lots of freedom, but also great responsibility. He now thinks the younger generation is more appreciative of their heritage.

> *I think they are more sensitive to an older building. They realize that it's unique since it's been here for a long, long time—and it's still standing.*

The Future of His Craft

Curtis doesn't feel his craft is truly in danger of dying out.

> *It's all in patience and bringing people up, but we keep doing it. I have three apprentices right now, so hopefully we can bring them up in here and just keep trying to build the company. My sons, of course, want to make the company bigger and grow and grow, which is fine.*
>
> *You know that is why I pulled away from my dad. Dad was in the latter stage of his life and wanted to retire. He just wanted six or seven people working with him. I wanted to be another Zachry* [a large San Antonio-based construction firm], *so I pulled out and went on my own. I told him I didn't want to do any more stone; I didn't want to do any more restoration; I wanted to go build H-E-Bs and schools and such. That's what I told him, and six, seven years later I'm doing stone and I'm doing restoration. I finally understood what he was talking about; I wasn't listening to him. I was just a kid.*

Curtis has not only fully grasped the significance of his father's message to him, but he is passing it on to the next generation.

> *First, there's an excellent market out there for you. Secondly, you get to work on these fantastic old buildings and there are very few people who get to do that; that's pretty neat in itself.*

Some Old and New Projects

Some of his most distant projects have been several of the ten courthouses he and his family have worked on throughout Texas.

> *Sonora is out there about 180 miles; Wharton is about 50 miles south of Houston; and then Corpus is about 140 miles down the road; Cuero, Beeville eighty miles each way; and then Atascosa.*

A pulley hanging from a scaffold used to raise stone, brick, or mortar to the work level.

Don Kirk

Michael Hunt at the Mission Espada job site. (Don Kirk)

We restored a building down in Pleasanton, the old rock schoolhouse for Louis Fischer.

His most distant work to date was to build a few fireplaces in Michigan. The Hunts also bid on new projects. They recently bid on a small medical building. Curtis, the businessman, notes:

We go after new, old, whatever's out there. We try to stay with restoration, but just to give my guys a break, we'll do a nice new project that's not quite as difficult as trying to restore old buildings.

Maintaining Heritage and Family Tradition

I'll stay with these old buildings and the beautiful forms they have. Just look at 'The Pearl.' I anticipate us being here three or four years. The owner's been real good to us. As long as we're making him happy, I think he will let us stay down here and work.

As far as down the road, it would be cool if my kids would work on something I did, or my grandkids. You know, something I built or restored. I had the opportunity to work on the exterior of the San Fernando Cathedral and my father did the interior in the 1970s. When we restored San Fernando, we took out this opening that my father had built in 1975 and, sure enough, Reymundo Rodriguez had signed the stone, and we found it and gave it to him.

(Don Kirk)

SAN FERNANDO CATHEDRAL

BUILT IN 1737, THIS GREEK REVIVAL FACADE WAS PLACED ON THE ORIGINAL SPANISH COLONIAL CHURCH IN THE MID NINETEENTH CENTURY.

HUNT, GARCÉS, HALL, AND PANKRATZ HAVE DONE RESTORATION WORK ON THIS JEWEL.

Working on the dome of the Bexar County Courthouse. In the center, Curtis Hunt; on the left, his brother Ross; and to the right, son Michael Hunt. (Courtesy Curtis Hunt)

Speaking of buildings seventy-five to one hundred years old and more, Hunt emphasizes:

> *You have to maintain them because they will deteriorate. That's basically what we do—come in and maintain them. Dig out the mortar joint, repoint them, keep the weather from destroying the masonry.*

"The whole charm of San Antonio is amazing; I go to other cities and I don't see it. I see a little bit in Austin, but it just seems like San Antonio's got so much more character to it."
—Curtis Hunt

We can rest assured, thanks to craftsmen like the Hunts, that our children and great-grandchildren will have a chance to appreciate the significance of the Alamo, be awed by the beauty of the San Fernando Cathedral, enjoy San Antonio's wonderful downtown River Walk, value the many Victorian courthouses in Texas, and admire the craftsmanship of many other historic buildings saved by these master stone masons. As long as the Hunt family is working, the stones will be strong, the mortar will be fast, the materials will be authentic, and the lines will be true.

7. Kurt Pankratz

(Monte Adams)

INTRODUCTION

KURT PANKRATZ keeps his workshop and his forge at a beautiful spot on the southern edge of the Hill Country near Boerne, a few miles from where he worked on a dairy farm. He remembers helping milk the cows twice a day as hard work. His farm chores, it turns out, prepared him for the long hours needed to practice a trade. Blacksmithing beckoned.

Working in one of the oldest trades on earth, the art of smithing, Pankratz can and does repair wagon wheels using an 1850s wagon-tire roller and tire shrinker. He uses ancient techniques and tools hundreds of years old when he does work for the Alamo and the missions. Pankratz's work is so meticulous and his new gate hinges for the Alamo so close to the original, he has to put his mark on them so they won't be taken for antiques.

He excels at modern techniques as well, constructing metal sculptures of his own design or building a copper fountain for a client's home.

Pankratz learned much of his dedication to excellence from his father, who emigrated from Germany in 1923, and is still active today in his nineties. For more than thirty years, Pankratz has supported his family with his extraordinary skills. "The only way we ever really survived is if we could fix everything," he says.

Kurt Pankratz.

Kurt Pankratz

BLACKSMITH

Kurt Pankratz was born and raised in a rural part of northwest San Antonio. His parents, George and Merva Pankratz, were loving and encouraging parents, allowing Kurt the space to grow freely, to explore, and to be creative, as long as good grades were maintained in school. Kurt developed a love of history and reading. Family outings often included picking up fieldstones to add onto the house they were building, because, during World War II, building materials were hard to come by.

Kurt often tells of his favorite Christmas present. His father arranged for a dump truck of salvaged lumber to be delivered just before Christmas so Kurt could build anything he wanted. It was to Kurt's delight. He built a treehouse, fort, wooden weapons, a raft, and many rabbit hutches.

Kurt's first job, at eight years of age, was working on a dairy not far from his home. During the summer he would ride a bicycle with coworkers from Bandera Road in San Antonio all the way to Comfort, Texas. Said Kurt, *"It was a treat to live and work in Comfort. It was almost like summer camp."*

At the age of fourteen, he became an electrician's apprentice. Kurt was proud to have had the opportunity to rewire his folks' home using wire, a breaker box, and receptacles salvaged from the job site. At fifteen, Kurt and his brother were hired by their parents to build a garage for the home out of fieldstones. Jobs like these for a child would be unheard of today, but the jobs Kurt held at this young age gave him self-confidence, a sense of independence and accomplishment.

While Kurt learned building skills working with his father, his mother wrote poetry, quilted, tatted, and made hairpin lace while working as an accountant. The family maintained a garden and raised rabbits for sale commercially. Kurt's parents were his example of dedication to excellence and hard work. It was often said in the home that "anything worth doing is worth doing well." From their example, Kurt's attitude for life was formed. He knew at a young age he wanted to work at building things by hand, and he also enjoyed designing. Understanding how mechanical items worked was always fascinating to Kurt. At twenty-seven, he left behind jobs at Southwest Research and Continental Airlines to fulfill his life's dream. In his mind was an idea to build unique items of wood and metal for clients with the primary goal of customer satisfaction. He loved antique machinery. It was what he could afford and could keep in operational condition. He rented a large grain barn from Lucille Ave in Leon Springs and in 1969 Kurt opened his business, "The Village Smithy."

> *Dad was happy that I was working at a trade. My parents supported my decision to start a business that I enjoyed. They had always appreciated my handwork with both wood and metal items.*

Kurt started his business in 1969, when he was twenty-seven years old.

> *I never knew any blacksmiths personally. I saw pictures and examples of blacksmithing. It was something that I was always interested in as a boy growing up.*

It was springtime in the Hill Country when we met Kurt Pankratz at his shop, The Village Smithy, in the middle of several acres covered with

Kurt's wrought-iron gate at Arneson River Theatre, San Antonio. (Tommy Adams)

Kurt's well-appointed office just outside his workshop. (Don Kirk)

oak-trees. The barn-like shop he built himself would appear as an enormous cavern if it weren't so full of equipment, tools, wagons, antiques, supplies, and samples of his work.

The Beginning

> *In the middle '60s, I decided to start my own business. Blacksmithing was in my background. I always thought there might be a need for more handmade products, and it would be unique if I could combine both wood and metals in my shop to make useful creations. It would be one place that architects and designers could go to find what they need fabricated. I have always considered myself a craftsman rather than just a blacksmith.*

From the beginning, Kurt endeavored to keep the hand-worked tradition alive, but he also wanted to put his individual stamp on his work and designs.

> *I wanted to try and do something I could be comfortable with and I also knew that I could help preserve, restore, and fabricate period items. I felt this was important to our history and culture and, of course, important to me.*

The blacksmith's most basic tools: anvil, hammer, anvil stakes, and tongs.

(Monte Adams)

Kurt's workshop plaque. (Don Kirk)

Self-Taught

I am basically self-taught. I also read a lot of books. When I was in high school, I did not get to take shop, something I feel every boy would enjoy. I had to find information where I could. There were a few old shops that I could visit. Johannes Scholze, [one of the master woodworking craftsmen featured in this book], *was a family friend. While in junior and high school, I looked forward to times we visited as Hans always had something to show me in woodworking. Hans was generous in sharing his time and knowledge. Whenever I had questions regarding finishes and carving, he was an invaluable resource. Most smiths and woodworking shops in the early '50s did not want to divulge their trade secrets; they were still trying to protect their jobs, and frankly, they were busy with their craft, trying to make a living.*

At the time there was only one other shop in San Antonio doing hand-forged decorative work, and one small business was still involved in the tooling processes and farm implement repair. The local blacksmiths were almost all gone except those in outlying areas, but the one in Devine, the ones in Poteet, and the one in New Braunfels were shut down. In fact, one of the anvils and several tools that I use, I bought from Schultz's Blacksmith Shop in Marion, Texas, back in the late '60s. They had already closed up shop.

"One thing you find out with a self-owned business is that it's a lot more hours in it than the average eight-hour-a-day job." —Kurt Pankratz

When I started my business, an old friend of the family, Ernst Reimer, a metalworker, told me I was crazy. He said that he couldn't keep his business going and he shut it down. I said, 'Well, I'm going to try.' After forty-five years I'm still here.

Kurt working the forge in his "barn." A forge has a forced-air source such as a bellows or blower to intensify the heat, an enclosure to hold the fire, and a chimney or vent to expel smoke and fumes. Kurt uses a charcoal fuel, but mineral oil, heating oil, diesel fuel, propane, or natural gas can also be used. (Monte Adams)

Below: Kurt's forge with coals, blower, and water bucket. (Don Kirk)

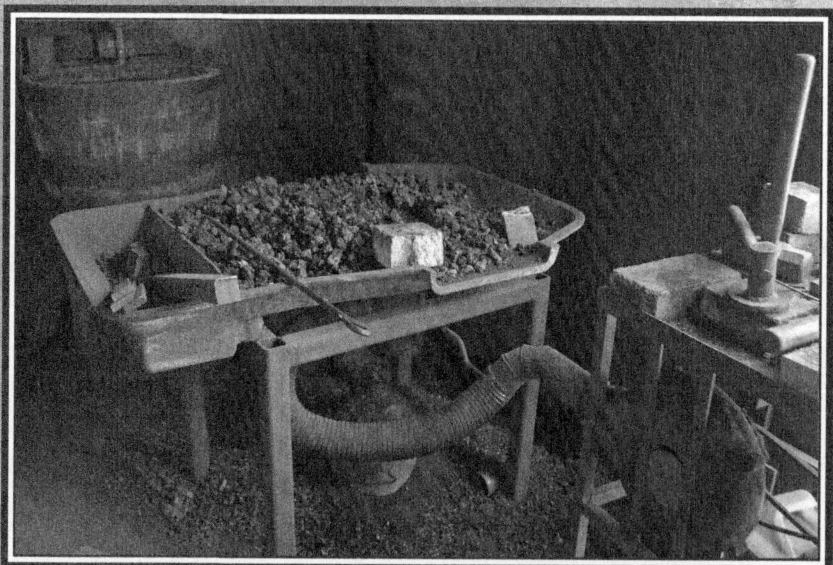

Above: Assistant Owen Zimmerman is using a blacksmith's nippers to construct a ring for a chandelier. Below: Toby Tobias is holding the steel ring as Kurt supervises. Kurt found Owen and Toby to be a great asset to his business. (Don Kirk)

(Monte Adams)

A SERIOUS VISE FOR SERIOUS WORK
WM' VANDERMAN, WILL' MANTIS CO. PATENTED JULY 7, 1897.

Apprentices

I've worked with a few apprentices who eventually started their own businesses, but none of them went into blacksmithing as a trade. I'm always willing to help others when they come to me with questions in hopes of helping preserve historical work done by old-world blacksmiths. We work with other shops if we don't have the machinery; it's not cost effective to have a piece of equipment that is seldom used. Shops do the same with us. Time becomes a huge factor when doing any project.

As far as teaching and training, I like to encourage anyone wanting to learn, but then it's also something that takes a lot of your time, and earning a living is one of the priorities in business. One thing you find out with a self-owned business is that it's a lot more hours in it than the average eight-hour-a-day job. Fortunately, I have George "Toby" Tobias, who has worked with me for twenty years, as I need him. Also, I call Owen Zimmerman when the project needs more hands. Owen has been with me since he was a young man, well over twenty-five years. Their skills have been invaluable to the successful completion of many of my projects. I could not have done without their help over these many years. Toby and Owen are excellent craftsmen. It is hard to find good help. I was very fortunate to have found both of them. They are outstanding in their work ethic, skills, and techniques, but mostly they are good lifelong friends.

One other person I often relied on was Howard Huntzinger. He was a master at hand lettering and gold leaf. He was an artist with an uncanny understanding of how signage should appear so that it is dimensionally correct. He helped me develop a sense of space in design. I was fortunate to have him as a resource on many projects.

A Family Business

Kurt's wife, Dana, works in the business, often helping with antique restoration and refinishing.

We have been married forty-five years. With her encouragement and support I was able to pursue the career of my choice. My wife now basically does all the paperwork, bookkeeping, and a lot of the legwork with errands. One daughter, Erika, enjoyed working in the shop and the other daughter, Lara, worked occasionally with designs. Our dog, Basha, a Pyrenees, is always walking with me to and from work. He is a constant companion and the official greeter to all my clients.

"Well, it is hot! The forge puts out a lot of heat. You're working with very hot pieces of metal, but it's something I have always enjoyed."
—Kurt Pankratz

Getting and keeping the steel at just the right temperature for bending.

(Monte Adams)

A wall of blacksmith tongs in every shape and size for every type of task. They must be able to withstand high temperatures. Types of tongs include straight, box jaw, knee, horseshoe, poppers, short jaw, U-bit, and wolf jaw. (Monte Adams)

Blacksmithing Has Been Around Forever

When men first started using metals to build things, they discovered that heating could change their shape. Most of the metal pieces—in early civilization—were for weapons. And, of course, they found out they could build tools, kitchen wares, and furniture out of various metals. Farriers learned to shoe horses and oxen to protect the animals' hooves. Different trades arose and specialized in metalworking, such as goldsmiths, silversmiths, tinsmiths, coppersmiths, wheelwrights, and coopers. Generally blacksmiths dealt with steel that was black in color.

Ornamentation evolved as blacksmiths discovered steel could be worked. It started with very low, very low quality malleable irons. Now we use steels because the old alloys are no longer in ready-to-use forms and have to be specially ordered. But, as they found they could work with steels with newer processes, it became a part of furniture, lighting, a lot of things.

The "Little Giant," a trip hammer, saves on shop time. Powered by an electric motor, the cam lifts a "hammer" repeatedly to flatten a piece of metal to a desired thickness.

(Monte Adams)

And what about blacksmithing today?

Mainly, we build light fixtures, furniture, kitchen utensils, kitchen racks for hanging utensils, fireplace accessories and screens—just about anything you can think of made out of metal and wood can be built in my shop.

"It's hot work, but I love it!" –Kurt Pankratz (Monte Adams)

Tools of the Trade

Like every craftsman interviewed, Kurt knows that tools are integral to his craft. He says the evolution of tools is also the history of blacksmithing.

Techniques have never really changed; the tools just became better.

> **"When I started my business, an old friend . . . a metalworker, told me I was crazy. He said that he couldn't keep his business going and he shut it down. I said, 'Well I'm going to try.' After forty-five years I'm still here.'"**
> **—Kurt Pankratz**

Kurt collected most of the tools he uses. Like many craftsmen, Kurt likes the old tools:

I've always felt that it was interesting to use tools that someone else had used. Often they are of better quality. They've had a history, and, if needed, I will repair the old machinery for my use.

Kurt will fashion his own tools if something special is needed. His shop is brimming with tools of the trade. He pointed to one.

The Trip Hammer

We acquired that trip hammer about forty years ago and it's a real help because we can use it for heavier pieces of metal. It takes a lot of the work out of just hand hammering. The trip hammer can cut down shop time. This particular one was built about 1880. It's called a twenty-five-pound hammer, but trip hammers can go up to hundreds of pounds of force.

A Champion Blower & Forge Co. Wagon-Tire Shrinker.
One of Kurt's favorite machines.

(Monte Adams)

Companies Still Making Tools for the Blacksmith

There are 'Champion' blowers and forges—probably one of the most popular brands. There is the 'Vulcan.' Another one in this country is 'Heller Tools,' which may still be in business. I'm not sure. Some of them have changed from blacksmithing tools into other lines of tools. You can see the crossover in some tool companies. My favorite: 'Champions' because they're more collectible.

Using What Tools You Can Get

Oh yes, a lot of tools we found were not working at all. We'd buy them and fix them up because that's all we could afford and was just what we needed. I have a thirty-six-inch band saw that's dated 1900. It came from Dittmar Lumber on Commerce Street when they were closing. It was not running, so I got it for scrap. I poured my own babbitt bearings to use in it, made some adjustments, and got it working. I've been using it for forty years and it will run another forty with the right care.

The Wagon-Tire Roller

The wagon-tire roller is one of Kurt's favorite pieces of machinery.

That piece is for rolling wagon tires, and that particular 'Champion' tool was built about 1850. We still use it quite a bit and it's still just in as perfect a shape as it ever was. It rolls the iron tires for wagons, that heavy iron band that goes on the outside of the wooden wheel, just like a rubber tire on a car.

The Wagon-Tire Shrinker

We have another piece that goes along with wheelwright work called the 'wagon tire-shrinker.' It upsets the steel wagon tire slightly while still on the wooden wagon wheel. It was quite an innovation for a blacksmith and saved him lots of time.

The wagon-tire shrinker I have is an interesting piece and in perfect condition. It dates back to the middle 1800s. This tire roller I found in an old shop in San Antonio that had closed up shop. The wagon-tire shrinker I bought from an old smith in Marion, Texas, when he was already retired. I don't think he ever used it. I think his dad, who was also a blacksmith, probably used it.

Cattle brands burned into an old barn door. (Don Kirk)

Modern Tools Save Time

While he his old tools, Kurt doesn't turn his back on modern tools when they are the best for the job.

> *For years my most-modern tool was my power hacksaw. It was a turn-of-the-century one that I rebuilt. I got it so I wouldn't have to hand cut large metal pieces. In the welding process, we use modern welding techniques. We can forge weld, but time-wise it's often prohibitive—it's an old technique, not within the budget generally, so we use the torch, TIG welding, and the wire-feed welder.*

Kurt does his forge work in the back corner of his shop with sunlight streaming through huge barn doors. But the season or the time of day doesn't come into play if there's a job to be done.

Terminology

Kurt was ready to explain some common blacksmith terms.

> *A 'hardy' is a chisel with a square shank that can be inserted into a square hole (a hardy hole) in a blacksmith's anvil. They are used for cutting metal and shaping. A 'fuller' is a blacksmiths hammer for grooving and spreading iron. 'Upsetting' is taking a piece of metal and drawing it together to make it shorter, in a sense. Then you have 'drawing,' the lengthening of a piece of metal rather than shortening it. There is 'twisting.' There are all types of heat treating and tempering and again 'drawing'—here the term is used differently. It means drawing 'temper' to the piece of metal.*

The Metals

Kurt says he does not have a favorite metal he prefers to work with.

> *As far as the actual smithing, you work with steel. Now we do some forgings with brass and copper. I do like the old low-carbon steels because you can easily work and shape them. I work with brass and copper also. The job determines the medium used.*

Sometimes great old metalwork is lost to history.

> *The old YMCA building they tore down had some nice work on it. Luckily, I was able to save a piece or two. They usually just scrap the work, which is sad, very sad. Well, I've got a couple of old pieces that came off the old YMCA building. One of the sign brackets was done without any welding other than using forge welding techniques, which indicated to me that the building must have been made sometime before the turn of the century. I am glad I have the sign bracket because of its excellent construction and design.*

"When there was a big craze,—I guess you'd call it—for all these trail rides years ago, we started repairing lots of wheels for horse-drawn wagons. Now most of the cowboys and their families ride in their air-conditioned motor homes and the horses travel in trailers." —Kurt Pankratz

(Don Kirk Photos)

Farrier and Wagon Work

Kurt said he never worked professionally at shoeing horses or mules.

> *We forged shoes for different farriers who had special shoe requirements—when they couldn't find one to fit from a catalog.*
>
> *The trail rides and chuck wagon cook-offs both had equipment that needed attention at times. The business goes in spurts. When there was a big craze—I guess you'd call it—for all these trail rides years ago, we started repairing lots of wheels for horse-drawn wagons. Now most of the cowboys and their families ride in their air-conditioned motor homes and the horses travel in trailers.*

It's Hot!

Kurt seems to enjoy the challenges of blacksmithing:

> *Well, it is hot! The forge puts out a lot of heat. You're working with very hot pieces of metal, but it's something I have always enjoyed. But I guess if you aren't careful you can sure burn yourself. Yes, I burned myself a few times; not very seriously, but it happens.*

The Art in the Craft

Kurt's artistry is most evident when he creates his own designs, which he is frequently called on to do. He has worked on projects for historical entities, churches, museums, the San Antonio missions, as well as numerous individual clients.

San Antonio Buildings and Special Projects

> *If you look carefully, San Antonio has a lot of early architecture that uses metalwork. The old library has some wonderful work on it. In fact, we restored the two* torchères *that stand in front of the old library* [now the Briscoe Western Art Museum] *a number of years ago. The metal had deteriorated and they were ready to topple. We completely restored them. We have had quite a bit of work in the King William historic district. Gates, fencing, iron cresting on roofs to be duplicated or repaired. We were pleased to work on the Walter Mathis home—a beautiful iron fence that needed missing parts fabricated and repaired.*

The Alamo

> *Hand-forged stanchions with chains were added to protect the front walls of the Alamo from the long lines of daily visitors. In the interior, we added grillwork so the original foundation of*

Iron stanchions in front of the Alamo and wooden Alamo bell stand fabricated by Kurt. (Courtesy Pankratz)

the Alamo could be seen by guests. Donation boxes and hinges also were needed to reflect the period. We designed with Wallace Saage, then the Alamo curator, a mesquite and iron bell stand for the Alamo's original alarm bell. The bell had been painted black and I was told it was made of cast iron. But no, I knew the bell had to be bronze if it dated from the Alamo period. After careful stripping and cleaning it was indeed bronze. I thought the project combining wood and metal turned out very nice and in keeping with the mission era.

The Missions

We did replications of gates at Mission San Joe. They were small passage gates. The granary doors at Mission San José were replaced. We replicated the clavos—from the one remaining original—to restore the doors to their original appearance. We replicated iron hardware that was incorporated with the original construction at Mission San Juan. We were pleased to be a part of a major restoration project at Mission Concepción recently, working with Father Jim Rutkowski. We refinished and restored the heavily painted altar to natural wood finish and fabricated a marble and polished steel Holy Water Font.

Branding irons forged by Kurt for use at the Alamo in living history demonstrations. The left one is the "Alamo de Parras," the right "Mission de Valero." (Don Kirk)

A processional cross for the seventeenth-century Corpus was fabricated. A moveable pulpit was embellished with copper using the pomegranate design on the front of the church. A Mission-style cabinet of mesquite and wrought iron was fabricated to conceal a mini refrigerator. The bracket holding the Holy Oil was restored and a matching bracket was fashioned for the Madonna on the other side of the altar. We fabricated donations boxes for several of the missions. I have felt privileged to fabricate work for the San Antonio missions. It was work near and dear to my heart—exactly why I started my business: to be involved in the preservation of history. The San Antonio missions are being considered for a designation as a World Heritage

The Alamo's original bronze alarm bell. (Courtesy Kurt Pankratz)

MISSION SAN JOSÉ, SAN ANTONIO, TEXAS
AUTHENTIC PERIOD RECREATIONS BY KURT PANKRATZ.

(Don Kirk Photos)

KURT'S EXACT COPIES OF THE ORIGINAL DOORS ON TH GRANARY. TH ORIGINALS AR PRESERVED IN HUMIDITY-CONTROLLED SPACE AT THE MISSION.

(Don Kirk Photos)

DECORATIVE ESCUTCHEONS WERE MADE OF THE SAME MATERIALS AS THE ORIGINALS.

"The fascination for architectural history is found in the details."
—Don Kirk

Kurt had to forge period hardware to restore this wood entrance gate at Mission San José, San Antonio, Texas. (Don Kirk)

WROUGHT-IRON WORK BY KURT PANKRATZ

Iron processional cross with Corpus.

Fabricated new bracket for Madonna.

(Don Kirk Photos)

Restored bracket and font for Holy Oil.

Fabricated Holy Water Font.

(Courtesy Kurt Pankratz)

MISSION CONCEPCIÓN, SAN ANTONIO, TEXAS.

The beautiful eighteenth-century Mission Concepción, San Antonio, Texas. (Don Kirk)

> *Site. The World Heritage designation would ensure the missions place in history.*

The missions received the World Heritage Site designation in 2015.

La Villita Historic Arts Village

Kurt also provided quite a bit of the hardware in the restoration of the historic La Villita in downtown San Antonio.

Bolt lock forged by Kurt, Mission San Juan. (Don Kirk)

> *It was interesting when my wife said that when she took back the box of hardware, both the original samples used to copy the shutter latches, hinges, and shutter dogs and the new hardware I fabricated,*

Kurt recreated period hardware for La Villita Historic Arts Village, San Antonio. Built by German immigrants, these buildings date back to the 1850's to 1880's. (Don Kirk)

they couldn't distinguish the old from the new. She had to point out that the new hardware bore my mark as the blacksmith. We were proud of that, as that's the intent of any replica.

For the La Villita Assembly Hall, we restored the original Lynn Ford light fixtures on the front of the building. We then duplicated additional hand-punched brass light fixtures for the River Walk area to match the original Lynn Ford brass cylinder light fixtures.

In the Arenson River Theatre area we provided folding crowd-control gates fabricated in iron as well as other gates and railings, to include the gate on 'Rosa's Bridge.'

Sullivan Carriage House

Working with Pape Construction we worked on the restoration of iron railings, fencing, and fittings for the Sullivan Carriage House. [San Antonio Botanical Garden]. *We repurposed the original iron pieces to fit into the new museum store and restaurant.* [See photos in Othon Garcés' chapter in this book.]

Institute of Texan Cultures

Kurt finds taking time to teach children a little history to be worthwhile.

> *Working with Jack Maguire, head of the Institute of Texan Cultures, I provided some hand-made items for several traveling education cases for elementary schools. The idea was to provide portable kits that could be checked out by teachers. They could then take the kits to their classrooms to show items of the Old West and Texas. This provided a hands-on experience for the children. The kits included items such as spurs with jingle bobs, cow bells, bridle bits, and belt buckles. Things that can make history come alive to all who are interested.*
>
> *I duplicated the Institute's logo on a branding iron. We branded the chuck box located in the Institute with the Institute's logo.*

One of several Lynn Ford bronze light fixtures reproduced by Kurt at the La Villita Assembly Hall. Below: Chapel stands. (Don Kirk)

Recreating That "Look"

Kurt sometimes uses the older methods himself, especially when needed.

> *The 'Thundering Hooves' was a nationwide traveling exhibit for museums on the 'History of the Horse' sponsored by the Witte Museum. Periods of history were depicted using full-size models of horses adorned correctly for their period. American Indian, Cowboy, Mexican Vaquero, and Spanish Con-*

A large steel kitchen rack with grape vines and candle holders.

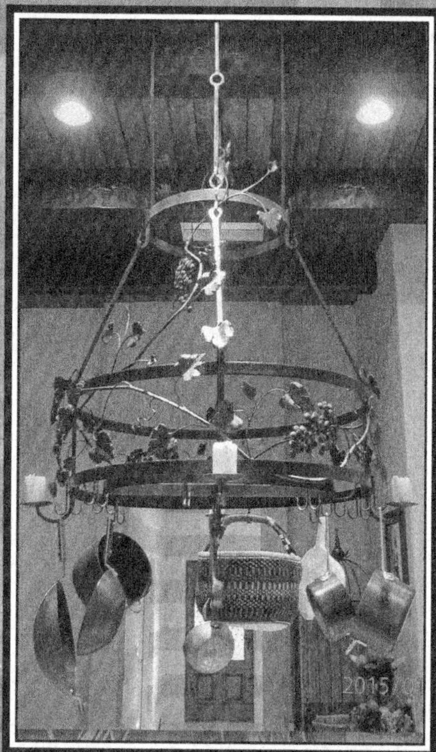

(Photos Courtesy Kurt Pankratz)

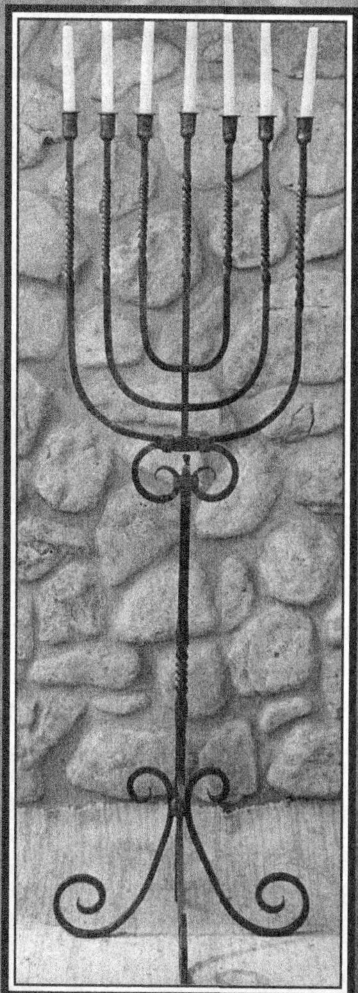

One of Kurt's first commissioned pieces, a forged-iron candelabra of twisted iron, 1968.

Kurt's beloved dog, Basha, a welcome fixture in the workshop. (Don Kirk)

An iron and brass dumbwaiter that lowers to a wine cellar.

A forged pendant light.

(Photos Courtesy Kurt Pankratz)

Beautiful forty-inch-tall, hand-forged steel fireplace andirons.

Old wrought-iron gate at Mission San José, crafted in 1935 by the WPA, restored by Kurt Pankratz. (Don Kirk).

Restored hand-forged lock set. (Don Kirk)

quistadors were represented. I replicated Spanish spurs, stirrups, bridle bits, buckles, and hocking knives copied from original pieces that were located safely in showcases in the exhibit. My reproductions were placed on the horse models out in the open. We fabricated all of the items exactly like the Spanish would have made them in the 1500s and the cowboy items in the 1800s.

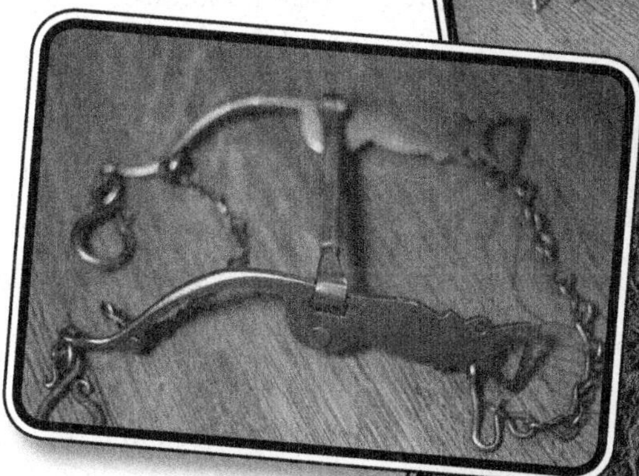

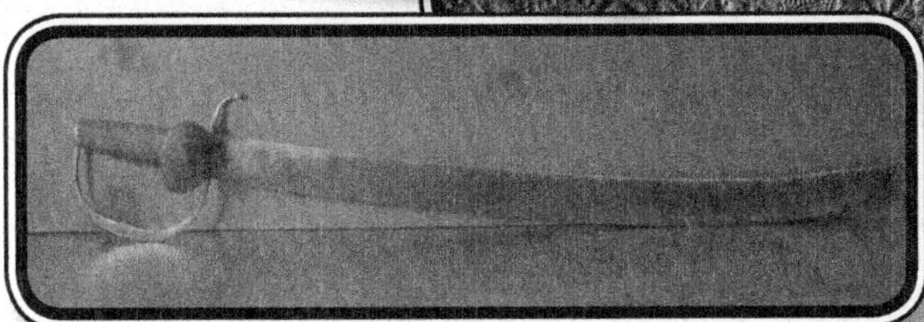

(Photos Courtesy Kurt Pankratz)

Making His Mark

Kurt uses a mark with his initials to stamp and identify his work, especially important when working with historical replications.

> *I designed a special mark, sometimes called a 'touch,' for my pieces as a signature for my work. I don't want it to be mistaken or be sold as an original.*

His Favorite Jobs

Kurt has been involved in many restoration and custom projects in San Antonio and the surrounding area. The projects have varied greatly and all were very interesting.

The Majestic Theatre needed iron stanchions to separate the seating in the balcony. Working with Milton Babbitt, we designed heavy twisted-iron pieces. The work pleased the architect and we were in keeping with the 1929 period design.

The City of Olmos Park requested I restore all their gas street lights. A few street lights we changed to solar to fill a grant they had received. My wife asked if I knew anything about solar. I said 'No, but would gladly learn.'

A beautiful outdoor electric light fabricated by Kurt using antique parts. (Don Kirk)

Our Lady of the Lake needed clock parts forged and fabricated for their beautiful 1920 tower clock. The best part of the job was getting to climb into the clock tower to see its inner workings.

The Transportation Museum in Galveston, formerly the Santa Fe Railroad Station, needed period lights for a new museum building.

The Karnes County Courthouse needed roof cresting fabricated. Fortunately, I did not have to install that job.

Larry Sheerin's pre–civil war cannon restoration was probably one of Kurt's favorite projects. After receiving copies of original working drawings from Watervliet Arsenal in New York, Kurt replaced all the wood for the caisson, fabricated missing iron pieces, replaced wheels, painted it with the original military colors, and fully restored it to operating condition.

Kinetic Concepts Inc. (KCI) needed modular conference tables. I designed rectangle, square, and semicircles that could be rearranged in various configurations. The tops were polished stainless steel banded with inlaid brass insets in oak and stained ebony black. The table bases were made out of oak stained ebony black with a stain finish. Mrs. Jim Leininger, the owner's wife, paid me

The restoration of this pre–Civil-War Mountain Howitzer required many hand-forged metal parts. *(Photos Courtesy Kurt Pankratz)*

AN 1850s CHAMPION WAGON-TIRE ROLLER

(Monte Adams)

the highest compliment. She said 'KCI had just ordered conference tables and what they received were works of art.'

The 'Souper Salad' restaurant chain presented an opportunity for working in aluminum. I fabricated, textured, and sealed six large aluminum sculptures for every store in their chain (about 160 stores as I recall) as well as sneeze shields. Mass production was not routine in my shop and it kept us busy for a couple of years.

The Hill Country Hyatt promotes local history at their resort. The hotel was built on the site of the Rogers-Wiseman Ranch. I crafted replicas of the ranch's working cattle branding irons and framed them for each suite. Each frame contained the two branding irons of the ranch with the ranch history for the Rogers-Wiseman Ranch. I also designed other items to fit their western theme. A miniature covered wagon for the golf course, old-fashioned pumps to clean golf balls, and bicycle racks of cedar, to name a few. It is good to see that large corporations such as Hyatt are sensitive to the preservation of our history.

Primarily Primates, a non-profit sanctuary for monkeys located in Leon Springs, needed three chimpanzee arenas for the chimps from the movie 'Project X.' That was a real challenge, entailing the use of twelve tons of steel for the eighteen-foot-tall arenas.

The chimps needed to be able to move from one cage to another but they had to be securely locked into their own cages for cleaning of the arenas. Safety and security was paramount, as they can be dangerous animals. My wife and I ended up donating time and working with the non-profit sanctuary for many years. We greatly enjoyed working with the animals and assisting in 'Primarily Primates' many building projects.

The San Antonio Zoo needed assistance when Mopie, their gorilla, figured out how to bypass the security of his cage door. I redesigned the locks on his door so they could be securely locked and could be tested. Mopie could no longer access the arena while the caretaker was cleaning. Gorillas and chimps are extremely cunning and smart. Locking mechanisms must be well designed and well thought out. They have nothing better to do than to plan escapes.

For Don Strange Catering and Rio Cibolo Ranch, I suggested a new use for a branding iron. I suggested they replicate the client's company logo in a branding iron. Hot-of-the-grill steaks could

Background Photo: Briscoe Western Art Museum torchères restored by Kurt Pankratz. (Don Kirk)

Kurt fabricated antique hardware to be used with an old paneled European door that was to remain unfinished to preserve its wonderful patina. (Photos Courtesy Kurt Pankratz)

then be branded, as well as making souvenir wooden plaques. The branding iron was proudly presented to the company. The branded steaks were a unique use of the irons and a reminder to the guests about Texas and its Old West history.

Diversity and Versatility

Kurt has remained in business for more than forty-five years through his versatility.

> *All my work has been by referral or word of mouth. We have never advertised. Projects I have worked on vary from repairing pewter steins, to the designing and fabrication of chimpanzee arenas requiring tons of steel. The projects have included a variety of metals and woods, sometimes combined in a work. Some projects have included the need for knowledge in plumbing and electrical wiring and were always supported by my desire to learn new skills.*

Kurt always has other projects to do when he doesn't have metalwork.

> *Right now the demand for hand-forged metalwork has become so great that there are companies that have developed machine tools that actually allow the mass production of hand-forged pieces, at least in appearance. They have the look, the feel, but to the trained eye, you can tell it's a mass-production piece. To stay in a customer's budget sometimes the mass produced items are used. I, however, prefer to make all my works by forging.*

Blacksmithing Will Live On

Kurt doesn't believe there is a danger of losing the blacksmith's craft.

> *I've seen a real upwards trend towards many of the crafts, and now there are blacksmith guilds all over the country. There are blacksmith conventions here and in Europe that I would have liked to attend, but it was just hard to get away for any length of time. Craftsmen are developing some wonderful, unique techniques. They have meets, conferences, trade ideas, and have workshops. In fact, I have just learned that Austin Community College has one of the largest blacksmithing schools in the country! I was pleasantly surprised. I would enjoy visiting the college one day.*

It is clear that Kurt Pankratz loves his blacksmith and woodworking shop. He loves the work, even if he has to open the barn doors for light, and he does not complain about the suffocating summer heat. Kurt knows there is still a lot of work to be done to preserve as much as possible of San Antonio's diverse and interesting history. He hopes this unique craft will

THE KURT PANKRATZ WORKSHOP IN A BARN.

A 1900s power draw saw. It draws the blade across the metal to make the cut. (Don Kirk Photos)

Farm implement wheels.

Forge and welding shop.

Tools and more tools.

Kurt on his workshop's front porch.

A wall of blacksmithing tongs, wrenches, and nippers. (Don Kirk Photos)

Front porch of Kurt's workshop.

Antique wagon wheel leaning on an 1850s Champion Wagon-Tire Shrinker.

Collections of farm implements, shop tools, and branding irons swamp the front porch of Kurt's metal fabrication workshop.

(Don Kirk Photos)

continue for generations to come. He feels deeply honored to have been selected Craftsman of the Year by the San Antonio Conservation Society and to be recognized in this book. It is the validation that Kurt has always strived for in his career.

And what is this?

Kurt would like to acknowledge some of his staunch supporters over the years. He offers a heartfelt thank-you for their support of his craft and their friendship.

> *I must thank O'Neil Ford, Gary Sprout, John Tarr, Ron Walsh, Orville Carr, Mid English, Jessica Stringer, Ken Bentley, Steve Tillotson, Tommy Adams. And then there is Betty and Bob Kelso, Ethel Runion, Dennis and Suzi Strauch, Dr. Leonard and Julie Rastrelli, Father Jim Rutkowski, Wallace Saage, Don Strange, Roemary Kowalski, Craig and Marjorie Christopher, John Rubiola, Carla and Mac Northington, David and Debbie Straus, Larry Sheerin, Lloyd Jary, Al and Nancy Perkins, Don Kirk, Steve Tello, David and Bonnie Brooks, and many more too numerous to list. Thank you all.*

A wall of Kurt's cattle brands burned into wood planks. (Don Kirk)

8. Richard Di Filippo & Family

(Monte Adams)

INTRODUCTION

VENICE ART TERRAZZO CO., a family-owned business founded eighty years ago in 1934, has successfully employed creative members from four generations of the Di Filippo family.

With each client, the Di Filippo family creates a customized, unique work, ranging from kitchen countertops to the expansive floor of an airport terminal or hospital lobby.

Artistry, creativity, and design are all needed for a satisfied customer. The years of experience of the Di Filippo family and Venice Art's workers ensure the style and elegance of marble and granite's cousin at one-third the cost. Terrazzo has a sophistication you just don't get in most other floors.

Modern terrazzo techniques began with fifteenth-century Venetian marble workers who used odd-sized marble pieces remaining from custom-made marble slabs to create surface terraces. The techniques evolved over time, resulting today in an almost infinite variety of designs available for this beautiful, durable, and virtually indestructible material.

One can hardly walk into an old building (and many new ones) in San Antonio without finding the Di Filippo family's work underfoot.

(Michael J. Smith)

LAWRENCE RICHARD DI FILIPPO SR. MARY RICK.

(Courtesy Di Filippo Family)

MATTHEW LAWRENCE MARY CATHRIN RICK.

Richard Di Filippo & Family

TERRAZZO ARTISANS

The word *terrazzo*, which means "terraces" in Italian, applies to both cement- and epoxy-based media mixed with aggregates, poured, set, and then polished into virtually indestructible surfaces.

This ancient craft—its roots date back five thousand years—is also a legacy for the future. Richard Di Filippo Sr., now the patriarch of the family, says that terrazzo

has a permanence and 'artisticness' to it that everybody sees and feels when they travel to Europe and see those old floors.

Amazingly, son Lawrence notes, "*the average person doesn't even know what terrazzo is,*" even though—from airports to hospitals to schools—they have walked, stood, run, and played on it for decades.

Thanks to the Di Filippo family, we learn the "inside story" of this beautiful material "hiding in plain sight." On an early spring day, several generations of Di Filippos shared their stories about the ancient art of terrazzo and their successful family business.

History of the Family Business

Richard Di Filippo Sr., flanked by sons Lawrence (a Vice President) and Richard "Rick" Jr. (a Vice President) and wife Mary (the company President) starts the conversation by telling us how his father, John Matthew Di Filippo, started the business.

The Early Days

> Mary: *John came down to Texas in 1928, and started his own terrazzo business in 1934. Originally, he came from Italy, settling in Minnesota and then in Texas. I believe he was eighteen when he left*

Italy and around twenty-eight when he left Minnesota. He didn't like the cold climate.

Richard: *I was born in Minnesota, but I only lived there one year.*

Mary: *I'm pure Texan.*

Mary met her future husband when she was only thirteen because her mother lived across the street from one of the people who worked at Venice Art.

Mary: *We've been married sixty years. I may be president now, but I didn't start here until my last son started first grade. I was supposed to be temporary, but I'm still here. I keep the books.*

The Italian Connection

Richard: *Dad* [John Matthew] *had a different trade in Italy. He was a carpenter. There's a little town called Spilimbergo near Venice—a lot of terrazzo people came from there and they started practicing the craft almost immediately upon their arrival in the United States. Dad was born in the town of San Daniele del Friuli, also in northern Italy. At the time, a large number of craft workers, especially the brick and masonry people, came from northern Italy. In fact, when they built monumental buildings like capitols, the people who imported the contractors over here imported all those craftsmen from Italy to do the masonry work.*

John Matthew Di Filippo, founder of Venice Art Terrazzo. (Courtesy Di Filippo Family)

My father was very close to that area in Italy, but he did a lot of carpentry. When he came to Minne-

The Terrazzo Men. (Courtesy Di Filippo Family)

sota, he first worked for a driving company, then he used his carpentry skills to build forms and to set the strips for the terrazzo. That's how he got into it initially—learning the trade there in Minnesota. Then he came down here and started the business in San Antonio, a branch office of Venice Art Marble Company. He was the manager. Then he bought the San Antonio branch and changed the name to 'Venice Art Terrazzo.' Since 1934, it's been Venice Art Terrazzo and in our family. And, we've been at this location pretty much since 1934.

One of my great-uncles, Henry Romani [John's brother-in-law], *and one of his best friends started the business with him. Eventually, they either sold out, retired, or just got out of the business, you know, old age, leaving Dad the sole owner.*

Lawrence: *Over the years my sister and all of my brothers—including my brother-in-law Richard Rasor—worked here growing up, but my brother Rick and I have chosen to carry on the business.*

Richard: *Even though it was 1934, Dad always told me he didn't even remember the depression, that it didn't affect him. It was just a hard struggle like starting up any business, you know.*

The office of Richard Di Filippo & Family. (Courtesy Di Filippo Family)

Wearing Many Hats

Richard Jr. continues, relating the day-to-day activities of a small family business.

> *Rick: I'm Richard Di Filippo, Jr., but I answer the phone as Rick.*
>
> Lawrence: *Rick and I are the estimators, and we work with the clients, but with a business as small and as complicated as this, you have to wear a lot of hats.*
>
> Rick: *One day you're talking to an architect and in the afternoon you're talking to the owners. Then in the late afternoon you're putting out bids. The next morning you're trying to make sure that all the billings are done and that all the men get out to the job sites. Sometimes, when they're all out on various jobs, material comes in on the truck and you have to go jump on the forklift and get it off the truck. You really do wear a lot of hats!*

MAKING TERRAZZO FROM START TO FINISH

Planning

Lawrence details the basic steps to planning a terrazzo project:

> *With terrazzo, we have various systems depending on the design or requirements of the project. Our available systems range from epoxy to cement, and from as thin as three-eights inch to as thick as three inches. This variety and ability to customize a system allows us to offer a product that will meet the needs of the designer or owner in a large number of projects and varying conditions.*
>
> *Generally, the first step is to determine the size and area where the terrazzo system is to be used. The next step is to determine whether there's a pattern, a logo, and or design element, as well as the number of colors to be installed. Then we work with the architect or owner in deciding on the best system along with a design mix in terms of the color, pattern, and aggregates. Once that process is done, then we make terrazzo samples for final approval before ordering material. Then, finally, the actual process of doing the floor work can begin.*

Prepping: Sanding and Strips

> *The first step in the actual installation is the floor prepping, or a sanding of the existing concrete or wood. Traditionally, we go over concrete. We can also go over a wood sub base, but it requires its own special preparation.*
>
> *Then there's the process of laying out the divider strips. The strips give us a place to stop our pour from day to day along with termination points for other materials. They can create a design pattern by themselves as well as divide the various colors. We lay out the strip pattern and then the colors are filled—sort of like a mosaic. You know, like a 'paint by numbers.' The different colors are then filled; each is hand-trowel applied.*

Mixing: Recipes for Texture, Color

> *It's a mixture of crushed aggregate, traditionally crushed marble. It can also be a blend of manufactured glass or recycled glass—and onyx and quartz, along with mother-of-pearl, and even things like a broken mirror. This shows you how unique each job can be. They're blended to a recipe in a percentage. For example, 25 percent of*

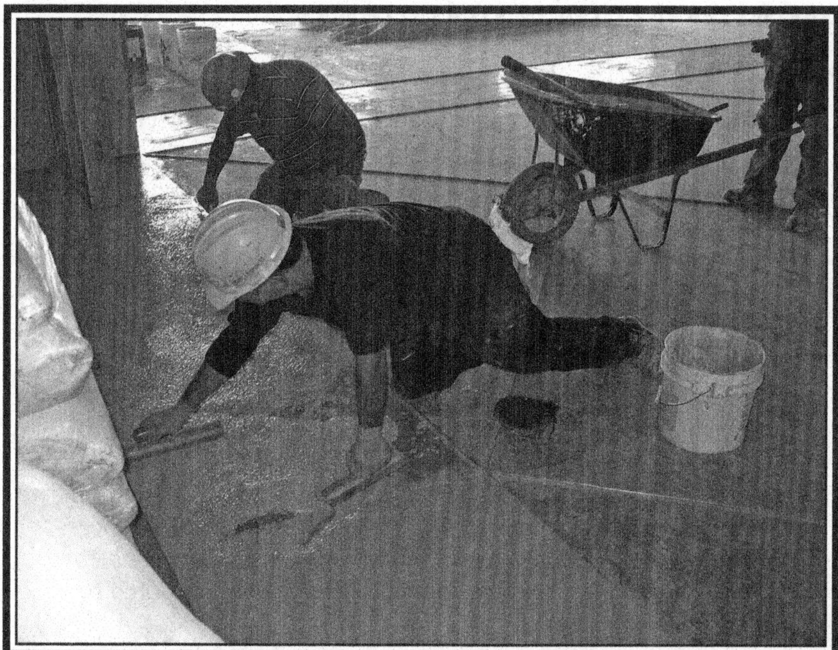

Spreading a crushed aggregate and marble concrete mix within divider strips used to separate colors or pours, CHRISTUS Santa Rosa Hospital. (Courtesy Di Filippo Family)

one color, 25 of another and 50 of yet another—making up a 100 percent mixture for each; creating its own unique recipe per color.

Then, you have the background color that we refer to as the matrix. Terrazzo can have a blend where you have aggregates that match the matrix or contrast with it. For example, you can get a red background with white chips, called aggregates, or a black background with white aggregates. Or you can have two different background colors, using the same chips or aggregate blend, like all white chips or all black chips. In other words, you can come up with an infinite number of recipes that are worked out first. Each design mix is indicated on the drawing, creating the pattern.

Application

You mix up the blended aggregate recipe and then mix it with the binder or the matrix (epoxy or cement) that holds together the aggregate. Then it's hand troweled within the strip pattern and leveled to the top.

Troweling the mix to bring it flush with the divider strips, Concord State Farm. (Courtesy Di Filippo Family)

In cement terrazzo, there is an additional process that involves a sprinkle of additional aggregate into the already placed floor mix. We 'hand seed' or 'loft' the aggregate into the mix to get a good uniform mixture. Then it's packed and rolled, creating a dense and compacted floor. This material sets up overnight.

Pouring and Setting Up

If it's a multicolored system, then it will involve multiple set-up times. Multiple colors can be poured in one day provided that two colors do not touch. Sometimes, you can pour even five or six, if it's a checkerboard pattern. If the job is large enough, then typically you pour all of one color for that day and allow it to set up overnight and then pour the next color and so forth.

Polishing

Once it is all set up, we begin the grinding and polishing steps. Before we had machines, the grinding phase was all hand worked. The machinery we use today incorporates resin bond silicon carbide stones or diamond stones.

Everyone has to work within the setting time of the concrete or epoxy mix. (Di Filippo Family)

Lawrence elaborates:

> *We start with a very rough, aggressive, 24 grit stone and then we take it up to 80, 120, 220 grit and so forth. We go from a rough grind to a polish. Like when sanding wood, you use sandpaper that goes from a rough to a fine grit. Gradually, the floor emerges. In the grinding process, you cut the divider strips, cut the aggregate, and then cut the entire surface. Then you actually get a cross section of what you laid down. That's basically how the mosaic comes from the pattern.*

The Strips

> *The strips stay there permanently. You can pick colors for them as well. They traditionally come in zinc, a silver color we call white metal, as well as an aluminum strip. We also have a gold color made out of brass. Today, there are even some plastics that are available in a variety of colors. You could even use wood.*
>
> *On some of the real old terrazzo jobs there weren't any divider strips; they used temporary wood strips and then removed them*

before pouring the next color. In other old installations, they put down little squares of marble cubes and those divided the colors. Those marble squares stayed in there as part of the final floor. You need to divide the colors somehow, whether it's temporary or permanent.

THE APPEAL OF TERRAZZO

An Ancient Heritage

Richard: *If you have ever been to Venice, in the Doge's Palace, the terrazzo in there is from the 1100s or something like that.*

Lawrence: *According to literature from the NTMA (National Terrazzo and Mosaic Association), they found what would be considered terrazzo dating to about five thousand years ago. It has been found in places like Turkey, Rome, and Greece.*

In addition to lots of pure mosaics in Italy, almost everything else they did over there was out of marble. So, when they carved it or cut it, they had a lot of scrap marble lying around. It did not take them long to figure out what to do

The Doge's Palace, Venice, Italy. (Don Kirk)

with those extra pieces of aggregate. The cut marble was in various sizes, and they continued to use those pieces until they were too small to carve or use as tile; then they crushed the remnants up further. They later mixed it with a binder and this created a new surface we now call 'terrazzo.'

The terrazzo elevator hallway and foyer for the Concord State Farm Building. (Di Filippo Family)

And the Italian terrazzo industry was born.

Its Lasting Nature

Today, terrazzo floors and other surfaces are not seen in residences as much as the were in the past. It used to be in the entryways of even the most modest of homes. It's mostly used in commercial applications, but even the type of commercial application has changed a lot.

It used to be schools and hospitals that used a lot of terrazzo, because they put it down in corridors and it lasts forever; it's nearly bulletproof due to its hardness; and it's easy to maintain. Schools—at least in the San Antonio area in recent years—have gotten away from the large usage of terrazzo. They use it in restrooms, the lobbies, and on the stairs because stairs take a lot of abuse and other finishes they've tried don't stay down very well. Terrazzo stays for a very long time. We just did 14,000 feet in a hospital in Westover Hills.

Appeal for Designers

Today, designers get creative with terrazzo. When it was in schools and hospitals in past years, architects liked it because it was easy to maintain and it went well with their glazed brick colors and tile colors. They didn't have to think about it too much. Then about fifteen or twenty years ago, designers started attaching themselves to the architects—and the designers liked using color and patterns. Now, it's gotten to the point where there are very few plain vanilla, one-color terrazzo jobs anymore; they're all quite intricate!

Aesthetic and Utilitarian Appeal

Richard: *It's a beautiful product, but it's also long lasting and durable. Architects, designers, and owners love it because of its unique ability to appeal to a wide variety of different people—whether they're interested in utilitarian, long-lasting floors or an intricate, decorative floor. All that affects the cost and the size. Generally, terrazzo is made for large open areas where you have lots of people walking about. You know terrazzo's going to hold up. So, that's why arenas, airports, courthouses, and schools use our products because they know that when you spend the money, and it's not unaffordable, it is going to be there a very long time and not need to be replaced.*

The "Green" Appeal

Since the terrazzo industry depends on rocks, you can safely assume it won't run out of that basic material anytime soon. And, as Lawrence points out, it's becoming an even more "green" industry:

It's the ultimate dream floor as it is: it is beautiful, it lasts a long time, and because it has a history of being made from recycled materials (marble chips) from the marble slab industry, they don't end up in a landfill. In the last five years, we have been using recycled glass, as well. As a country, we need to conserve and recycle. We can do something with all the recycled glass they're picking up. Industry crushes it up and uses some of it as decorative landscape material, but also we're actually using it as aggregate in our terrazzo floors!

MAJOR RESTORATION PROJECTS

Richard: *The opportunity for historic jobs goes in cycles. Right now, there's a real interest in historic preservation—courthouses and state schools—but that percentage of our work is usually low.*

A lot of courthouses are being restored. We're going to restore the one in Goldthwaite, Texas, near Lampasas. This particular terrazzo application doesn't have any divider strips. It still has border and field colors, just no strips. Also, I just did some pricing on the one in Trinity County, which is in Groveton. We're also working on one in Cuero. Those are all old terrazzo applications.

Lawrence making samples of different color options. (Courtesy Di Filippo Family)

The Color-Matching Challenge

Lawrence: *We're not buying a pre-manufactured product from somewhere that limits us on our color options. If we don't get it right on the first sample, we make another one. Most of the time for a project we'll make probably an average of four samples.*

Richard: *You get a piece close to where you're going to make the repair and you start with that.*

The State Capitol

Lawrence: *As far as the aggregate, we have imported a number of North American and a wide variety of Texas chips. With the state capitol project, there were early Texas chips originally used, some of them not readily available today, if at all, and so we were trying to go back and match them.*

Richard: *When we restored the capitol, we made around a hundred samples. There were nine different colors throughout, but they were installed at different times. When we use a cement terrazzo system, there's a dry color dye or a pigment that you use to color the cement. The cement itself comes in either white or gray. You can blend the two as well. So, you may use half white, half gray and then add the color to the mixed cement or add a color to just the white or gray by themselves. You've got to start doing that to create a variable color palette.*

The investigative process into historical buildings becomes an even more complicated process:

We were trying to go back to the quarry. We were saying, 'Well, this chip goes back to the Texas Quarry. It started around sixty or seventy years ago. The capitol was built much earlier, but when they put in that terrazzo, it was during a remodeling seventy-something years ago.' You have to go back and try to find an aggregate that was used at the time. Investigate what's produced then, versus what's produced now.

It's a natural product, so the black seam that's in the quarry now might be dark black or a little more gray, or have a white vein in it, or have some other contaminants. It's quite an arduous process to try to find colors today that will look like the ones that were used then, especially if you can't actually find the original.

A team of Terrazzo men on the job. (Courtesy Di Filippo Family)

When we did the state capitol, we worked closely with the Bilbrough Marble Company. Together we actually went to some closed-down sections that they had worked years ago. They went in there and got some special aggregate for me so that we could get a better match.

TRICKS OF THE TRADE

Trade Secrets Are No More

Terrazzo artisans have their individualized signature "recipes" for color and sometimes techniques. The family patriarch, Richard Sr., points to an early photograph on the wall:

See that picture up there, the one with the Italian workers? All those guys standing around, that's how they used to grind terrazzo before they had the machines we do now. But, when the men first came from Italy, they wanted to keep the trade a secret. They built walls around where they were going to work. They wouldn't let anybody watch how they did it.

Any number of colors and patterns are possible, Golden & Drenner. (Di Filippo Family)

Rick: *You can't get away with that now, because if the architect doesn't know about terrazzo, and he needs terrazzo in a project, he calls you up and says 'educate me on terrazzo.' You give him all the information he needs so that he can put it in the flooring specs. The project then goes out for bids, and then we can get the chance to bid the job. Then, he turns around and wants you to give him submittals on how you're going to do the job, so that he knows you're qualified to do it—even though you gave him all the information in the first place.*

Rick shakes his head.

Materials Standards

Richard speaks to the standards:

First of all, the marble that you get has to be a certain hardness. In the old days, you couldn't grind it if it got too hard, so it had to be a particular hardness—not below ten on an abrasion test, I believe.

The terrazzo industry has an association, NTMA, that the majority of terrazzo contractors belong to. The association was formed by the contractors around the 1930s. In the late '20s, early'30s, the contractors got together. They are still talking about some of the same issues over these eighty-some-odd years. Whenever they would meet annually, they'd come up with standards. They tested the hardness of the different rocks, and they found out which ones worked better.

Need for Records and Privacy of Formulas

Lawrence: *As far as the blend, if we're working with the owner or the architect, we're working with the percentages that they would like to see. We keep that formula for future reference. It is private to a point, but we obviously share that with the people we are working with, and sometimes it is on record in the specifications. Many times they're not that concerned about it because they know we're going to do the work.*

Sometimes, though, we will get calls from our competition that want to know the formula, and we <u>do</u> share that information with them. It's just prudent that you have a system to keep track, because it's limitless in what you can do. If someone comes in and says, I like that, you have to know how to duplicate it.

Creating a counter top of confetti-colored glass chips for Palo Alto College. Below: The finished counter top. (Courtesy Di Filippo Family)

A Recent Example

Lawrence: *Right now, there's some jobs that Dad worked on years ago. When Rick and I are making repairs on some of these, we go to him for that information.*

Richard: *Right, just like that renovation of the State School in Austin that I worked on originally.*

Lawrence: *Dad worked on that, and their contractor had to do some repairs—remodel the bathrooms to make them ADA compliant, remove doors and whatnot. So there's a hole on the floor that we have to patch. The contractor called me up and sent me a picture via email. He wanted to know if that was something we could use to make a sample. I said, 'Well, it's a lot better if I have an original piece of floor,' but he said he didn't want to cut any of the floors out.*

A Tracking System

Richard Jr.: *We used to, when we developed a sample, put a sticker on the back and write the formula on it. Then when the job was done we would put the samples on the board. Well, it's a very dynamic room; architects come in and take a number of samples with them. Sometimes they come back to exchange them for others, and a lot of times they keep them for use on other projects. And, there go our samples! The formula is on the back, but thankfully, it's also in the job file. If the sample walks off as in a recent job, they call me up and say 'Okay, out of the four colors, I like this color from the Meadville Elementary School', and that's all they give you. So, a few years ago—probably over ten years ago now—we realized our system was not working too well. We have a numbering system now. Every floor sample we make has a number and then a reference number we keep in a notebook <u>and</u> in the computer.*

Mary: *Our files go back to 1960 when Richard, Sr. started in the office. The job Richard was just talking about was done in 1964 so we have a record.*

Richard: *I'm making a couple of samples based on those formulas. We're going to see how close they are; it gives us a start-up range. Today, there are a lot more formulas in the blend than there ever were before—because, in the past, it was simpler, more standard.*

The 6th-floor logo in green, blue, and yellow at the University Hospital downtown San Antonio. Below: In the entrance lobby of the Freeman Coliseum, San Antonio, Texas, the men are grinding down the epoxy (or concrete) to expose the marble aggregate of the new design they are working on. (Courtesy Di Filippo Family)

Sprinkling marble chips on the central circle of the design. Below: The new terrazzo floor in all its beauty and durability. (Courtesy Di Filippo Family)

TERRAZZO BASICS

Sources of Supplies

Rick talks about supply sources from the beginning of their company to today's available sources:

> *We used to get some aggregate from Italy because the ships would come to Houston. In time, that went away.*
>
> Lawrence: *Then, we began to get the majority from Texas quarries.*
>
> Rick: *The quarries in Marble Falls have been there since I can remember; it's the reason the town got its name. The way they do it at the local quarry is, if you're interested in marble, they'll lease land from ranchers or landowners and go in there and quarry the land. Then they will bring it to the crushers and they will crush it up for us, to the size we need.*
>
> Lawrence: *Areas around the country produce certain colors. We are fortunate in Texas because they produce a wide variety of colors. Georgia produces a white, a black comes out of Virginia, and some other aggregates now come out of Washington State. There are also Canadian chips and now México is bringing up onyx and different things.*

Most of the material used by Venice Art today comes from North America but may actually originate elsewhere. Rick explains:

> *There is a supply company in Chicago that <u>does</u> import Italian chips—some greens and reds—and now they're also bringing in chips from Turkey and China.*

Size of Pieces

Terrazzo is made of rocks from the earth that are quarried and crushed up into needed sizes.

> Rick: *Terrazzo is based on contrast to light and particular sizes are important.*
>
> Lawrence: *Each marble crushes in a different way, some a sliver, some a cube, but basically they all have to pass through a screen of a certain size so that they are all uniform in size. What we call a 'number one' is an eighth to a quarter of an inch. A 'number two' size is a quarter to three-eighths. We even have a smaller size, 'zero,' and larger sizes—all the way up to an inch and a half, but we don't use that size very much.*

Rick Di Filippo putting a sealer on test samples to show the client. (Courtesy Di Filippo Family)

> *Other countries are now discovering the world market as far as aggregates are concerned. We get emails all the time from supply houses bringing in granite from Brazil and from Africa, as well as Turkey and China.*

With something as heavy as stone as a raw material, transportation is a real issue. Rick notes:

> *We're fortunate that we have some major quarries within 150 to 200 miles of us. We're buying it right from Texas quarries, which produce terrazzo for a lot of the rest of the country as well.*

Skilled Labor Issues

> Lawrence: *You can't just go hire somebody and put them to work right away. They have to train; it's a learned skill. We train them ourselves.*
>
> Richard: *Currently, we probably have about eighteen to twenty people, but there's been way more at times. It's up and down. During my dad's time, it was sometimes close to 100, but then again it was more labor intensive back then.*

TERRAZZO MANUFACTURING TECHNIQUES

Cement Terrazzo Versus Epoxy Terrazzo

Richard: *Terrazzo was originally a cement or concrete material holding together the marble. Then, in the '50s came the polyesters and the '60s, epoxy. Now the whole terrazzo industry is probably more epoxy terrazzo now. Epoxy is a liquid resin–based material that holds together the same aggregate, but with it you can go thinner and you can get the pigmenting exact. In other words, you would be hard put to get the blue color of the Texas flag with cement terrazzo. With epoxy you can get that exact shade. The aggregate with the epoxy can be colored more than the cement.*

Lawrence: *Yes, the epoxy matrix can be tinted like paint, so there's a wider variety of colors that you can get; whereas, with cement you start with white or gray. White is the standard. White cement is more consistent than gray and we can control the dry pigment better. Gray cement is anything that is not white, so there's a lot of variance in gray even though some gray is greener than others, some gray is grayer than others, and some are browner.*

Rick: *That's the thing about terrazzo, with cement or epoxy, you have colored the background and then you added the color chips. Sometimes, you want a red chip with a red background, or blue chip with a blue background, or sometimes a contrast. When you think of color with terrazzo, you've got to ask: 'Am I going to go subtle or am I going to go bold so the chips stand out?'*

Patching with Epoxy

Richard: *If we're going to do a new floor—like the new floor in the courthouse in Goldthwaite—they want it to match the original look.*

If I'm going to patch quarter-size holes in the floor where they moved toilet partitions, or an eight-inch hole where they moved the toilet, it's easier for us to use epoxy, especially if we're going to patch a crack, because cement doesn't have a lot of strength when it's thin or the crack is real small, whereas epoxy does. Epoxy also matches better than cement.

TOOLS

Basic Tools of the Industry

Richard Sr. speaks about how the tools have changed through the years:

> *For the most part, we use the same hand tools as the old-time journeymen—similar to the trowels that bricklayers use. What has changed, is the electric grinding machines and the use of diamond grinding stones.*

A terrazzo floor-grinding machine. (Monte Adams)

Sanding in the Past

Richard asks his sons,

> *'You know the best place for sanding is standing upright?'*

> Rick: *Obviously the work was more labor intensive; in some ways it was much harder than it is today; it certainly took more time to trowel and work flat because this would help the grinders out. The mechanics would trowel, roll, and sweep the floor to remove some of the slurry on top, creating a smoother floor that would take less sanding or grinding.*

Early Grinding Tools Versus Electric Grinders

> Richard: *Today we use an electric floor-grinding machine and it weighs probably six to eight hundred pounds. It has disks underneath it with stones that grind. It's more aggressive than pushing stone by hand.*

> *They laid the floor a little tighter back then so they wouldn't have to grind as much. They didn't have the machines that are available now. You don't have to spend as much time working the floor today, knowing that the machine is going to take off more and save some of the hard handwork.*

"The average person doesn't even know what terrazzo is." —Lawrence Di Filippo

e Adams)

MACHINE &
CO INC.
OLIS MINN.

It was a big deal when electric grinding machines were first introduced. We have an advertisement brochure from 1927; it has one of Dad's uncles in it, and they were advertising the new electric-powered floor grinding machine. So, before 1927, they were grinding by hand. And this new machine they were advertising showed grinding. They were on a job and they were grinding two hundred thousand square feet of terrazzo. Imagine doing that all by hand! All those guys in that brochure had big smiles on their faces.

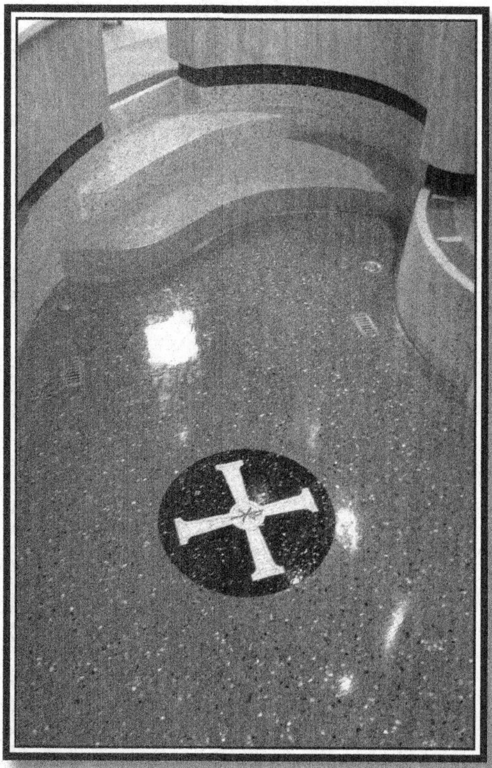

St. Matthews Baptismal Font.
(Courtesy Di Filippo Family)

Lawrence: *Some of the other changes have been going from pumice stone, a hard abrasive, to carborundum, to now diamonds, so it's changed quite a bit.*

BUSINESS OF TERRAZZO

The Competition

Richard: *As far as competition in the terrazzo industry, you will find that most of the terrazzo companies are third and fourth generation and they're just big Italian families. The family businesses are becoming less and less now. They're being sold and other people are moving into the market; they're a different type of people. In Texas, we have more contractors than probably most states. There are four terrazzo companies in Houston, two in Dallas and several in the Valley. There once were three in San Antonio, but there's only one here now—us.*

When a terrazzo project is bid, everyone bids it, but, over the years, we've been fortunate to have a good reputation for quality work at a competitive price.

Cementville terrazzo bar top prepared in sections and installed below.
(Courtesy Di Filippo Family)

Colorful terrazzo hallway at Randolph Brooks Federal Credit Union. (Di Filippo Family)

A Complicated Business

Lawrence adds:

> *One way to think about it is to think of it as part of the flooring industry, you know, wood, carpet, terrazzo, and so on. In a way, terrazzo is a small part of it because there are hundreds of tile contractors, hundreds of carpet contractors, hundreds of vinyl tile installers. It's amazing to me that everybody installing all the other products tend to judge their standards against terrazzo, in other words, trying to look like terrazzo even though we're a small little industry compared to them.*
>
> *It's also amazing that the average person doesn't even know what terrazzo is. Probably most people have walked on, stood on, and seen terrazzo and didn't even realize what it was. They think it's tile or something else, because it's something seen in commercials on TV or that you walk on in a hospital. Amazing! Just amazing!*
>
> Rick: *Most of the commercial floors—the carpet, the vinyl, and the tiles—those are all manufactured products that someone just*

installs. But terrazzo, because we're coming up with the recipe—mixing the aggregate and the binder—we're actually manufacturing the product at the job site.

Cyclical Nature of the Business

Rick: *This year is a good year; last year was a bad year.*

Lawrence: *No, last year was a good year.*

Mary: *No, it was the year before that.*

Rick: *No, it was 2007!*

Richard explains the reason for the cyclical nature—if the economy and the building industry in general are down, terrazzo is down, too:

Terrazzo is a finishing material; you don't build a building with terrazzo; you finish the building with terrazzo. So, if building projects are down, terrazzo is down. Of course, there's usually a lag between when they slow down on buildings and when our work slows down because finishes are not at the front end of the building, they're on the back end, so, there's usually a lag between slowdowns and when construction picks back up.

In the early '70s, real estate was a huge, huge market, and they were just building buildings with the intent to occupy them for a couple of years and then sell them to somebody else. They didn't necessarily need a permanent floor because they knew five years down the road they were going to sell to somebody else who might want to change the finish of the building. So, in the early '70s, the terrazzo business was way down.

Current Di Filippo Business

Most of the current Di Filippo business is new commercial work.

Lawrence: *Public areas.*

Rick: *Museums, courthouses, lobbies, entryways, banks.*

Mary: *Now, people use terrazzo for countertops and that's starting to flourish. Ours are different than the granite and other products a retail outlet like Home Depot has to offer. It's custom work, whatever people want, whatever colors they want, whatever size they want. We did one that was a kidney bean shape, I mean, you can't find <u>that</u> on a shelf!*

Aztec calendar terrazzo, Aztec Theatre, San Antonio River Walk entrance. (Courtesy Di Filippo Family)

THE CRAFT OF TERRAZZO

The roots of modern terrazzo are in Italy, Lawrence says.

> *Almost everyone—the majority of the terrazzo artisans—have a second or third generation that can be traced back to a town in Italy. Every ten years, we return to Italy with our industry for a convention!*

Family Roots

Mary notes the reason her husband enjoys going back to Italy:

> *It's his father's homeland. Richard finds people there that are the same as his father. Where his mother came from—in Luca—the wife of one of her cousins was still living there. Back then we stayed with them whenever we'd go back. And he saw the house there that his grandfather was born in—in northern Italy above Venice. His father owned the house until 1964.*
>
> Richard: *I would have liked to have bought the house, but we had five kids and couldn't afford it at the time. I have dual citizenship.*
>
> Rick: *We're all eligible for dual citizenship, even though we sons haven't applied for it because Dad's father came to America and*

was naturalized here. Dad was born before his father was a naturalized U.S. citizen so because of that fact, he was . . .

Richard: *Because of that, we're eligible for dual citizenship and I wanted that. I can get around the other traveling people out there, flash that dual citizenship! They're envious of that!*

Learning the Trade

Rick: *It's a trade that you have to learn.*

Lawrence: *You don't jump in with cold feet.*

Mary: *You have to work with somebody.*

Rick: *That depends on the individual. Some are very handy and quick learners and some . . . are not.*

Lawrence: *It would take several years for somebody to go through all the different steps, the processes, and really learn them.*

Richard: *And then there are degrees of workmanship that you see in a lot of the people you have working for you. They're not all experts, but they are all good workers.*

Lawrence: *Some workers are better at grinding and others better at troweling.*

A detail at the Aztec reveals that anything is possible in terrazzo. (Di Filippo Family)

Courses to Study

Richard: *Some chemistry courses probably would help. I took courses at reading architectural blueprints.*

Lawrence*: I went to Texas A&M to get a business degree, because I had planned on working here. I took some geology classes, because you're working with rocks, and marble, and minerals. There were some architecture classes that were helpful in reading plans and specifications as well as being able to converse with an architect. You have to speak their language.*

FUTURE OF TERRAZZO

The Next Generation

Lawrence: *During past summers, Rick's and my children would work on the jobs and in the office. It was exciting for us to see them work in the field just as we did. This experience has allowed them to get a hands-on feel while learning the craft. Having the fourth generation learning the business is very exciting. Its something that their great-grandfather and grandfather would be proud of.*

Currently my son Matthew Di Filippo and Rick's daughter Cathrin Di Filippo are working with us in the business now. They both graduated from Texas A&M, Matthew with a construction science degree and Cathrin with a human resources degree. My youngest son, Andrew, is also thinking about working here as well. They are bringing in a new energy and skill level that is up to date with the latest technologies. It's encouraging to know that this business has the potential to continue on for many more years.

The Di Filippo Difference

The Di Filippos do not sign their work, although Richard Sr. recalls:

We used to do work for a general contractor, Charles Madden, he's dead now, and he used to have a little plaque with his company name. Every time we'd do a job for him he'd put it in the corner somehow. He was proud of his work.

It seems that every house and commercial building ought to have a cornerstone, like in the 'olden days,' with the names of the owner, designer, and contractor—maybe more people would take pride in their work and build better products. The Di Filippos clearly take pride in theirs.

Richard Sr. muses about the unique nature of the Di Filippo craft:

It is very client dependent.

The 2013 NTMA Award Winner at the Randolph Federal Credit Union ASC II; an intricate fifty-four-color, sixteen-foot-diameter emblem. (Courtesy Di Filippo Family)

Lawrence: *We've been contacted time and time again by owners and they say, 'Well, we've inquired around and you guys are the ones to go to if you need new work or need something repaired.' In other words, any terrazzo guy can make a quick repair, but we take a lot of pride in the quality of our work.*

While there is no singular Di Filippo style, you can depend on this fourth generation company for quality materials, custom designs, attention to detail, client satisfaction, and, oh yes, works of beauty, durability, and permanence. Venice Art Terrazzo is carrying an ancient heritage of creativity and skill into a welcoming future.

Detail of Credit Union emblem. (Courtesy Di Filippo Family)

The Randolph Brooks Federal Credit Union ASC II emblem.
The 2013 Honor Award Winner from the
National Terrazzo & Mosaic Association.
(Courtesy Di Filippo Family)

PEOPLE HELPING PEOPLE ★ RBFCU ★ PEOPLE HELPING PEOPLE

9. John Hall

(Montre Adams)

INTRODUCTION

JOHN HALL, owner of Bluegrass Inc., does business as Hallmark Cabinetry and Millwork. He grew up in Louisville, Kentucky, and first came to San Antonio to attend Trinity University. He graduated with two degrees, one in business administration and one in economics, but he chose a career in building. His attention to detail and his fine craftsmanship have kept him in demand. His work is represented at the Alamo, the San Antonio Museum of Art, San Fernando Cathedral, and the Atascosa County Courthouse, among many other significant buildings in South Texas. The San Antonio Conservation Society has recognized John with the Lynn Ford Craftsman award for his lifelong achievements in historical preservation.

For John, work on historical buildings often requires historical tools. Some date back as far as 1860 and have names like four-head moulders, louver groover, and sash sticker—all of which he repairs and keeps working. They help him recreate the functional and elegant woodwork of the past and create stunning new pieces that will be prized antiques long into the future.

(Monte Adams)

John Hall.

John Hall

WOODWORKER

John Hall, located in the same building on Melrose Place since 1976, started with a much smaller space. The growth of his business over the years has been organic.

> *I believe the more space you have, the more you are going to use it. I think that I have expanded one, two, three, four, five, six fold in this building as my interests have changed. I've even expanded into the building nextdoor for my metalwork.*

John and his wife, Byrdy Wagner, have two sons, Cabell and Garrett, who have grown up at the shop. Early on, this was their day-care location. Speaking fondly of those times, Hall notes:

> *There was never any problem with hammers, saws, and nails in their hands. They never hurt themselves. Later I would supply their kindergarten classrooms with little boards and nails and all the kids loved that construction play area.*

When asked whether his children understand all that he's done to preserve many of San Antonio's famous historical buildings, he answers with his usual humor:

> *Well, if I tell them enough times, they might remember it—all young folks seem to have a short attention span!*

It is likely that most of us will recognize and remember the quality and the beauty of John's work. It is simple, subtle, and stunning. John Hall, in short, is a maestro of woodworking, a cabinetmaker without equal in South Texas. Here you will learn a little of how and why he came to share his gifts with us.

One cool spring day, John Hall sat down to tell us how and why he became a cabinetmaker:

> *Even with those two degrees, in business and economics, I didn't want to wear a coat and tie every day. I got a job working as a carpenter. At first the jobs were small, but then I moved*

on to larger jobs on historical houses across San Antonio. Backed with this work experience, I eventually became an independent contractor.

As a contractor of commercial as well as residential work, I found that I needed custom items not readily available. It was then I started my own cabinet shop. I realized I had to be versatile and that I needed special equipment to offer a complete line of custom-made items. That started my long search for vintage machinery that could help me accomplish my goals.

Starting Out

John recalls when he was about seven or eight years old:

My father put a dormer on the back of our house and he and one of my uncles would do different woodworking things. My father, William Henshaw Hall Sr., worked at a couple of jobs that had shops like welding and fabrication. That kind of got me interested in welding early on and I knew I wanted to make things.

We didn't have the games they have now when I was growing up. I don't think we ever had enough money to afford the new mechanical toys, but we did have nails and hammers and stuff like that we could play with. We'd go down to my dad's shop and make a mess,

John Hall matching a milled piece to a full-scale pattern. (Monte Adams)

but didn't get scolded too often. You know, I learned by taking things apart to see how they work. I think it's good for the kids. I have encouraged my sons to just go figure it out. I am not one to give them the easy answers.

I wanted to build things in wood and metal and it's taken me quite a while to get back to metal, and it is funny and strange to remember that my first big tool acquisition was a welder when I was just starting work after college. I have been involved in the wood business for quite a while. It suits my personality and my desire to control how the project ends.

John in his shop, December 2014. (Battersby)

Although he has always had competition from other woodworking shops in San Antonio, John's craftsmanship, skills, and attention to detail keep him in demand.

There are other cabinet shops, but not many that do the variety of work we do, from restorations to matching existing pieces to even creating new pieces from a photograph.

When the San Antonio economy declined in the 1980s, he left remodeling and focused strictly on woodworking, with projects like restoration work at the Alamo.

The Alamo and Other Historic and Public Buildings

I started working with people who were renovating the Alamo. We completed custom windows and the Long Barracks display cases, including the diorama display in the chapel at the Alamo. After that I completed several exhibition venues at the Museum

The Casa Rio restaurant on the San Antonio River Walk, identified by a train of colorful umbrellas. (Don Kirk)

These original doors were custom made for Casa Rio by John Hall. (Courtesy John Hall)

of Art, including the "Splendors of México" exhibit. My work is in many churches, from the Basilica of the National Shrine of the Little Flower to First Presbyterian, Christ Episcopal, and the Chapel of the Incarnate Word, along with the Bishop Jones Center for the Episcopal Diocese. If you count universities, then I have pieces at St. Mary's University and Trinity University.

We have millwork throughout downtown San Antonio. All the doors along the River Walk at Casa Rio are ours. The doors on the Eckert's down on the River Walk and up on the street level are ours. Three doors on the street level at Schilo's Delicatessen are ours. The Bexar County Courthouse. The list goes on.

The pulpit was lowered and new stairs built by John Hall in 2002 for the San Fernando Cathedral, San Antonio. (Courtesy John Hall)

At San Fernando Cathedral, John reworked the pulpit built in the 1880s. When the cathedral was restored, he moved the pulpit, lowered it, and rebuilt the stairs to it.

For the Atascosa County Courthouse restoration in Jourdanton, Texas, John worked on all the exterior windows and doors and many of the interior doors to meet the Texas Historical Commission specifications. The exterior windows are nine feet by nine feet and made using insulated glass to comply with current building codes.

John Hall's shop office and fascinating museum. (Monte Adams)

AUSTIN HARDWOODS, LTD
San Antonio
2002 AUGUST 2002
2002 SEPTEMBER 2002
2002 OCTOBER 2002

We did that work using old longleaf yellow pine beams salvaged from a building built around 1880 in South Carolina. The beams were fourteen inches by eighteen inches square by twenty-six feet long. They were milled down, and the windows and doors were made from that stock.

On the historical side of the work, I've been pleased and blessed that I've been able to be in San Antonio. I have worked all over town at most of the universities and many of the significant buildings that have needed restoration. It has just been wonderful to be asked and able to do these things.

When I am doing [a historical piece] *you have your head down, and you are just doing that one piece and it may take you a day, or a week, or a month, or nine months. It's not something that you go out and have your name on, but I do enjoy knowing that I have pieces around town. I'd much rather work on historical buildings than new ones. The new ones don't offer the challenges of old buildings.*

The first large window job I did was the Hotel Havana in San Antonio. For this job, I went around to all the glass people and

John restored the doors and windows of the *Casa Navarro State Historic Site* in San Antonio. (Courtesy John Hall)

The restoration of the 1914 Mediterranean Revival-style Hotel Havana in San Antonio required new window sashes and some doorjambs. (Don Kirk)

asked whom the glazers were and found we have lost many of our glazers that rework old wooden windows. So, we had to build the windows and then learn how to glaze ourselves.

Luckily, my father gave me a few tips. At one point in his life he did remodeling. After a tornado in Louisville, he spent six months glazing windows. He gave me a lot of good tips and in turn, I have taught quite a few people how to glaze. So glazing is the art of putting the glass in place and having something to hold it in place. Humidity and temperature have a lot to do with successful glazing. It's a lot like making candy; the weather needs to be right. This is another artful detail in finishing the project and one of the final details to be proud of.

John Hall's Motto:
"Hand sanded; when we're done, you're finished."

John with new windows setting up in a glue press. (Courtesy John Hall)

Historical Tools

> *The tool—no matter where it is—is only as good as the person behind it. I have a bunch of tools that date back to the 1860s and they are a lot more experienced than I am. I try to keep them busy.*

John speaks of his old woodworking tools as if they are alive:

> *They are experienced, and they know what they are doing. The old tools work together in sequence because they are of the same period and made to work together.*
>
> *They do have their own quirks. Each one has something that you have to go and deal with on a pretty regular basis. The oldest machines, more so, are flat belt driven, which is a big mechanical belt that runs on the pulleys, which wear out more often than the new ones that are direct drive. You have to pamper them a little bit more. The old ones you have to oil every day. Like people and pets, it is always something that you have to deal with.*

Stories of how he acquired these amazing tools are tales in themselves:

> *I met different gentlemen in the trade. The first one I met was John Sievers* [J.H. Sievers Altar Manufacturing Co.] *and he and his father and his grandfather made church furniture in San Antonio to be shipped all across the southern states.*

Next, I picked up some tools from Joe Telford, who had a store fixture business. Over the years, I bought tools locally from Superior Millwork; others came out of Dallas and I brought some down from Kentucky. I have tools that date from the 1860s to the 1870s, even the 1890s. I'd say the average age of our tools is about 1930.

The 'moderns' are about twenty years old. You can create with any of the tools; you just have to be a little bit more aware of the antique tools. They have a little more play in them, not tight like some of the new ones. They're definitely not as fast, but most of what we do is 'one-off piecing.' I mean that we set up and make just one unique piece and don't do another like it.

So the boys with the big, more modern machines, if they do one piece, they have to stop everything to make it. In the end, they're not much faster than what I can do with my old equipment.

Like most craftsmen, John has both a thrifty side and an emotional connection to the people who made the older tools.

The price of my equipment was right because most people don't like the old equipment. Everybody wants the new fast machine with the computers and all that. I have those machines too. It's not the computer that makes the machine work, it's the person

Form with wood screw clamps for building arched windows. (Monte Adams)

JOHN HALL HARD AT WORK IN HIS SHOP IN 2014.

(Battersby Photo)

"The tool—no matter where it is—is only as good as the person behind it. I have a bunch of tools that date back to the 1860s and they are a lot more experienced than I am, so I try to keep them busy." —John Hall

1871 Bentel, Margedant, & Co. Universal Wood-Worker.

Phil Hall.

Sash sticker detail.

1940s-era jointer.

Man and Machine—Machine and Man.

(Monte Adams Photos)

Four-Head Moulder.

Drill Press.

(Don Kirk Photos)

Sash Sticker.

Louver Cutter.

John Hall.

(Battersby Photo)

John turning a column on his lathe. (Battersby)

that makes the machine perform. I was more interested in having the tool and the capability than having the burden of going to the bank. Don't forget, I built this business from scratch. There are big differences between the old and new machines, some obvious, some not. Old equipment is cast iron, heavier and slower.

The old machines can weigh from 8,000 to 10,000 pounds.

In some machines, the babbitt bearings added considerably to the weight. After ball bearings were invented, the old machines became a little lighter. There are advantages for custom work in what some see as drawbacks in the old machines. Slowness makes it easier to control the product. Our custom work demands accuracy and patience.

John notes new machines are more expensive, faster, and more lightweight. He says new machines can move and often do not give the straight, true cuts of the older, heavier machines when cutting thousands of feet of molding. Old planers run at 3400 RPM and cut to 1/32 of an inch tolerance, while new planers run at 7000 RPM.

He provides an example of working with his old machines—two four-head moulders, one built in the 1920s and the other in the 1930s.

> *There is a different knife for each of four sides of the wood. The knives can be set to cut a profile or a straight cut. Knives must be ground and set properly by hand, then aligned so all four work with one another to produce the right product. I have a large collection of knives to work with and sometimes we might find a match, but often we will have to start over and regrind in order to match a certain detail. That is what I have become known for.*

New porch columns replicating the originals. (Courtesy John Hall)

"Custom work demands accuracy and patience." —John Hall

With newer machines, says John, you need to do a huge volume to make the machines profitable. He specializes in precision and historical replication, not volume, so the old machines serve him better.

His wide-belt sander was one of the first in San Antonio (1954).

> *It came originally from England and I acquired it from John Sievers. A tool manufacturer representative from England visited me and said he had never seen that model before. This sander typifies another true advantage of my older machines—reliability.*

Shop desk, plans, and phone book to track down needed lumber and hardware—and to connect them to people. (Don Kirk)

These plans call for double-hung windows and cabinets. (Don Kirk)

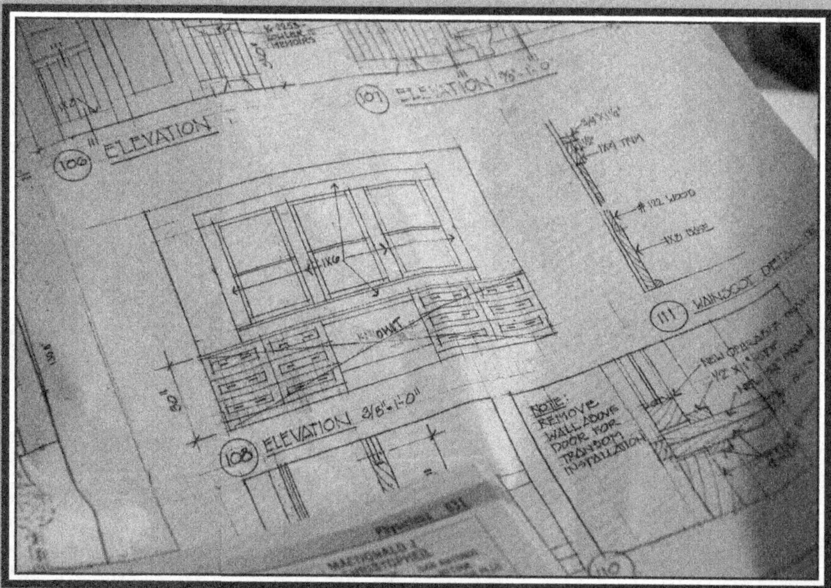

And he still uses the Bentel, Margedant & Co. machine that was built in November, 1886. John also acquired this machine from the J.H. Sievers Altar Manufacturing Co.

> *It's a horizontal 'shaper-joiner' and is used to make flat arches over windows and doors. It can also make single-sided molding. The bed moves in several planes: up, down, in and out.*

Tool Maintenance

> *I am the person who fixes most everything. People ask me where I get the parts for these machines. If it's an old-style babbitt bearing, then I have to pour that. If some of the cast-iron pieces break, I have a specialty machinist braze or weld those pieces. I am the one to put them back on. For the most part I am the one who sets them up and maintains them. It is a love/hate relationship sometimes.*

> *I am a 'tool man.' I know the workings of each one of these machines, even the newest CNC and specialty lathes that I use to make the balusters and porch posts. My life has been one of knowing my tools and how they can work to create something of great quality from a plain block of wood.*

On Wood

John tries to find the same wood as the original structure when doing a restoration. Often the historical projects cite "reclaimed clear longleaf pine." He has to be a wood "sleuth" to find a source for the longleaf pine. He also uses mahogany or Spanish cedar for restorations if no particular kind of wood is called for.

> *I don't think I have a favorite wood. Every wood has a different way it works, and they are all used in different parts of furniture and fixtures and windows as architectural elements. You have to look at the wood and know what you're going to do with it.*

Finding Old Woods and Old Hardware

> *There are old carpenters, and there is old wood, so you can find things if you ask around enough. I really like to connect the people with the things. Recently, the old Joske's building was being repurposed, and when I read that, I just knew that there was an opportunity to match that particular out-of-town developer with a local wood reseller. Today that matchmaking helped with the salvage of those old Joske timbers to be repurposed to the restoration of the Bexar County Courthouse. No, I did not get a*

Bronze Indian statue, stolen and eventually returned to the dome in 1987. (Don Kirk)

New arched windows rebuilt by John Hall. (Don Kirk)

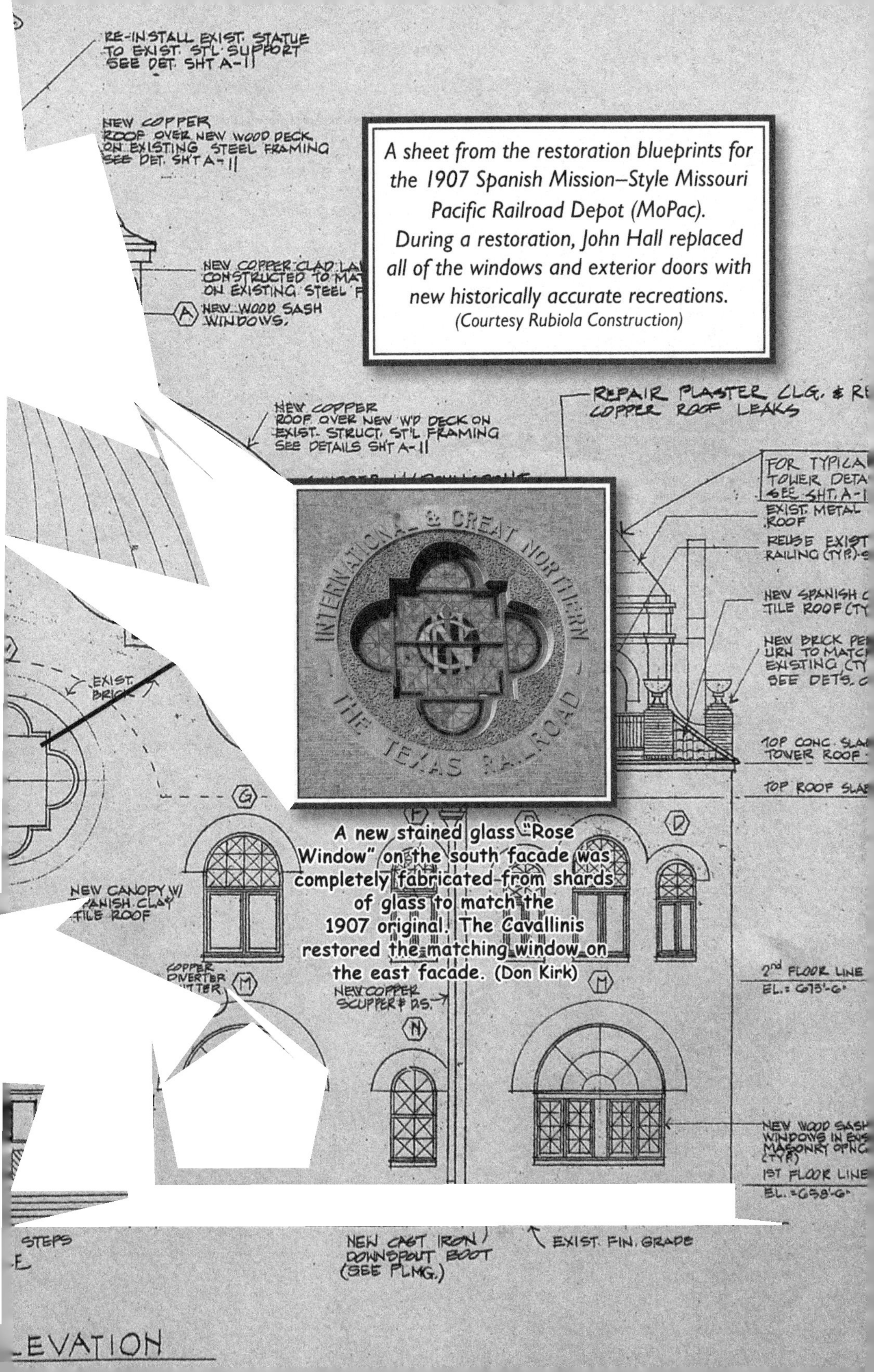

A sheet from the restoration blueprints for the 1907 Spanish Mission–Style Missouri Pacific Railroad Depot (MoPac). During a restoration, John Hall replaced all of the windows and exterior doors with new historically accurate recreations. (Courtesy Rubiola Construction)

A new stained glass "Rose Window" on the south facade was completely fabricated from shards of glass to match the 1907 original. The Cavallinis restored the matching window on the east facade. (Don Kirk)

The dilapidated 1907 Missouri Pacific Passenger Depot at "Cattleman's Square" in 1980. Called the "Taj Mahal" by its Architect, Harvey L. Page, the copper dome had been stripped by scavengers, with broken stained glass and rotten windows adding to the decline. Page's design was influenced by San Antonio's Spanish missions, particularly the San José. On top of the eighty-eighty-foot-high dome stands an Indian posed with an arrow pointed toward the northeast, in the direction of San Pedro Springs. Below: The Missouri Pacific Depot in 2014, after restoration with new windows built by John Hall and the east stained glass window removed and restored by Cavallini. VIA Metropolitan Transit has plans to make it a major Amtrak and commuter station for San Antonio. (Don Kirk Photos)

piece of that specific restoration work. I just linked those people together, something that I have learned to do. It's like matching hardware. I don't find much in San Antonio so I will call all over the United States because San Antonio is a user and not a manufacturer for a lot of these items.

Before and after restoration of an arched window on the Sullivan Carriage House. (Don Kirk)

Evolving His Craft

John does not have a typical project nor any typical days. He says,

> *Today, after forty-two years, every project comes by referral.*
>
> *I have found that I have become one of the old timers, in that often I am called when the project has a lot of repair to match existing pieces or is just so unique that I'm able to figure how to do it and make everyone happy. Never the low bidder; but safe to say, I make it right to everyone's satisfaction.*

Apprenticing

> *I never apprenticed. I learned by the seat of my pants. I learned from figuring out the machines and looking at the pieces that are being duplicated. I did have some old timers and other craftsmen that I worked with who would answer a few questions. But for the most part, I picked up the machines and had to figure out how to make them work.*
>
> *It takes investigation to figure out how old timers did it. Which piece went first, second, and third. It involves working and*

John Hall about to test for a snug fit on his mortise and tenon joint. (Monte Adams)

Some tough-to-make "multi-light" window screens for an old residence. (Don Kirk)

thinking in different planes. So, for most of what I've done, I've learned by myself with help from other woodworkers.

Over the years, I have had many gentlemen who did stints working with me. I recall fondly my brother Phil, and other guys, too many to count, but the stories I could tell! Perhaps if you asked them they may recall that this was like an apprentice shop. Never doing the same thing from one day to the next.

Future of His Craft and Craftsmen

John has serious doubts about the future of his craft.

I can't find people who want to work the craft. For sure, not everyone can deal with this cantankerous old equipment or even the new dang computerized ones. Most guys want a big paycheck on Friday, and it's not something you get easily until you have a lot of years behind you.

I think it is important to have craftsmen who have specialty trades. I think we are losing too many—not just craftsmen—we are losing all types of real workers in our country. I know there is a place for everybody to

Many windows were restored or rebuilt on this, the original classroom and dormitory of the Ursuline Academy, a convent established in 1851, now the Southwest School of Art. Many of the craftsmen in this book worked on this historic property. (Don Kirk)

come and have a niche. I have a very small piece of historical work and it sustains several of us.

I hope that when I retire somebody will want these tools that I have and carry on the craft. I don't know who is going to do it. Our school systems have decided that it is not necessary to have handwork in schools anymore. So, we have such a limited number of hands-on training programs in our schools—woodworking shops, auto shops, homemaking shops. I know we need to turn that around.

There are still wonderful tradesmen. My thinking is that the future is just about people, and pride in a good finished product. That is what has driven me to control the quality of everything I do.

Future of Building Restoration

When asked about the future of buildings that might need restoration fifty years from now, John shook his head.

I'd say it's pretty bleak. I don't know what quality of craftsmanship will be available in the future. They may even have 3-D laser printers to craft some parts down the road. I just believe that

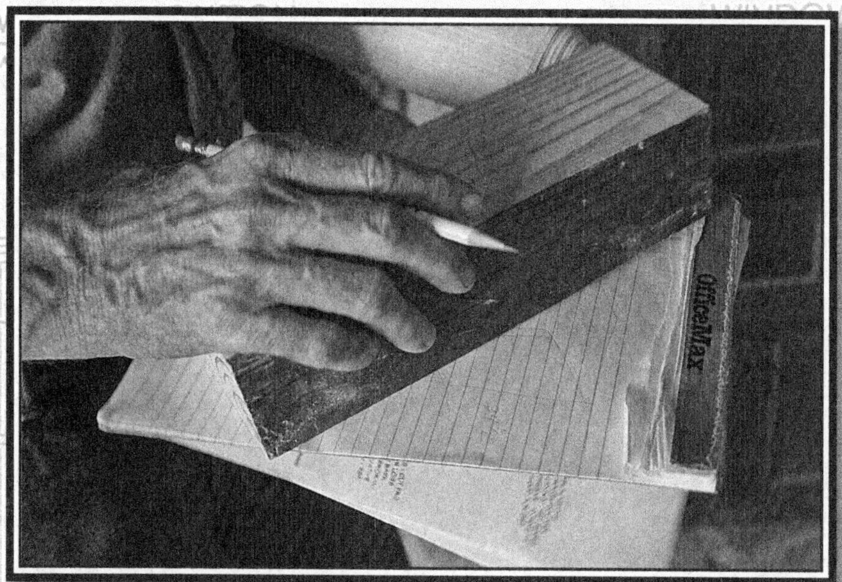

Fine, expert craftsmanship starts with the hands. (Courtesy John Hall)

this world will have the good sense to continue with the old way of crafting. It takes money to take care of historic buildings to preserve our past. The lowest bidder does not always deliver the best quality; their work suffices, but is not historically accurate.

For example, we are doing windows right now. We have to look at some of the windows and doors and we're either going to restore them, or we're going to make them match the style of the historical building—in this particular case—in the 1930s. The newer techniques of woodwork don't lend themselves to the old styles. It is very difficult to tell the new machines how to do the old ways. I can get some of this equipment to do it. My 'sash sticker' that was built in the 1890s is specifically built for making window sashes. This one machine does what three different machines can do.

Handcrafting and Manufacturing

The craft element of what we do actually goes hand-in-hand with the type of pieces we make. If we are doing architectural pieces, as in doors and windows, there are big manufacturers that can do those and pump out a lot of them.

When you get down to doing one unique piece, or you have to take one and repair it, well, now you need a craftsman. You need some-

body with the skills to go and work on that one piece and decide how it needs to be joined, how to take it apart. In a lot of instances, you have to take the piece apart in order to make the other part of it and then put it back together. The craftsman has the understanding of how it all fits together. I have great tools here, but, as I said earlier, the tool is only as good as the person behind it.

Working with His Hands, Working with Machines

John's hands have a symbiotic relationship with his machines.

I enjoy working with my hands. I was that like as a kid. When my father's lawn mower stopped working, I asked him if I could take it apart. Well, I don't think it ever got back together. But I did take my mother's oven door apart and it did get back together—and quickly. Since then, I've liked to work on and work with machines.

It's one thing to be able to walk in and have the machine spit a piece out. It's another thing to go in and work the machine so you can bring it back to life to make it do what you want. That's my craft.

I like to do a little bit of everything. My wife, Byrdy, says 'a John of all trades.' I do a little electrical. I do plumbing. I do work on the machines. I do woodwork. I weld. I am an auto mechanic. I built our house. I built most of this shop. I pretty much keep busy with my hands. I even do the dishes.

John's workshop: a lumber yard and mill under the same roof. (Don Kirk)

Atascosa County Courthouse

Drawing for Atascosa County Courthouse windows. (Courtesy John Rubiola) (Photo: Don Kirk)

(Monte Adams)

All four elevations are identical, with many windows that had to be rebuilt for

this Mission Revival–style Atascosa County Courthouse, Jourdanton, Texas. (Don Kirk).

Restored entrance doors, First Presbyterian Church, San Antonio. (John Hall)

Quality

John lives his own philosophy, and it shows in his work.

> *Quality is something that everybody has a different opinion of. I don't want to make anything that I wouldn't want to have at home in my house. So consequently, we take a little bit more time to try and get it done right. I've had the men a little upset with me when I tell them to go back and do it again and that sort of thing, but, I think in the long run, it has helped me in everything I do.*
>
> *Whether a historical courthouse renovation, or the homeowner's repair and/or replacement of the staircase railing, we try not to have any recalls. I'm not the only one who takes such pride. It takes quite a few of us to make something happen, and when everybody is doing a good job it is shipped out and everyone is proud and happy.*

Words of Wisdom

As John finished talking with us, we asked for words of wisdom. He replied in words echoed by all the craftsmen in this book:

I think if you want to do whatever your craft is that you just have to go and put in a lot of time and long hours. Look at the piece that you are doing and don't necessarily try to cut corners. Try to make the best piece you can.

You don't have to do it full time to be an excellent craftsman. I know some other professionals who do excellent work in wood as a hobby. To be a first-rate craftsman you must put something personal in what you do.

I love my work and find it interesting enough to still be doing it after forty years. I will probably be doing it until I can't get out to work anymore. I'll always be doing something with my hands.

John Hall on Fay & Egan Tenoner.
(Monte Adams)

Fortunately for us and future generations, the work of John Hall's hands—and his tools—have contributed immensely to the preservation of historic buildings, giving them new life and vigor, and John should be proud of his artful craftsmanship in a time-honored profession.

A pair of entrance doors built for Trinity University's Board of Trustees Conference Room, restored by John Hall.
(Courtesy John Hall)

10. Manlio & Adrian

Cavallini & Sons

INTRODUCTION

THE CAVALLINI COMPANY stained glass designs can be found throughout the Southwest as well as in churches in the Virgin Islands, Nuevo Laredo, and Italy. The company is especially known for its religious art, large-scale architectural installations, and historical restorations. One of their more recent and challenging installations—incorporating design, leaded glass, and mosaic—can be seen at Gallery Nord in Castle Hills, Texas.

Some of their restoration work in San Antonio may be viewed at Sunset Station and San Fernando Cathedral. Recently, The Cavallini Company installed protection glass on windows in St. Anthony Cathedral Basilica in Beaumont, Texas, that withstood the winds of Hurricane Rita. Commitment to quality is ingrained in every Cavallini employee—from Manlio and Adrian Cavallini, who now runs the day-to-day operations, to the artists, designers, and fabricators. The Cavallinis now compete successfully statewide and throughout the Southwest for the most prestigious commissions. If you can envision it, the Cavallinis can create it.

(Don Kirk)

Manlio Cavallini.

Manlio & Adrian Cavallini & Sons

Leaded, Faceted, & Etched Glass Artisans

Manlio Cavallini, a native of Rome, Italy, graduated in 1948 with a bachelors degree from Pontifical Gregorian University in Rome. He came to San Antonio in October 1949 and attended St. John's Seminary. In September 1950, he enrolled in St. Mary's University. In January 1952, he married his wife of sixty-two years, Viola Irvin, who passed away in 2014. He graduated from St. Mary's in May of '52, majoring in philosophy. From 1952 to 1957 Manlio Cavallini worked from 3:00 p.m. to 11:00 p.m. as a bottle inspector on the evening shift at the Lone Star Brewery while working to build the Cavallini Brothers Venetian Mosaic Company.

Manlio and his brother Publio started the mosaic company in 1953. Publio worked with an Italian company in México City by the name of *Mosaicos Italianos*. The company worked with very famous Mexican artists like Diego Rivera, Miguel Covarrubias, David Siqueiros, Juan O'Gorman, and Carlos Mérida.

Manlio described how his brother specialized in mosaic installation while working for this company and his participation in some of the projects created by those artists. For example, he took part in the installation of one of the beautiful mosaics at the University of México City. Then he decided to come to this country, and talked Manlio into starting the mosaic work.

For several years it was rough going until he and his brother decided to go separate ways–Publio pursuing tile contracting and Manlio continuing with the mosaic work. Manlio's company was incorporated in October 1961 under the name of The Cavallini Company, Inc. Soon after, the company expanded to include leaded glass, and later faceted, beveled, and etched glass—offering design, fabrication, and installation services. In 1972, Publio Cavallini decided to move to San Diego, California where he established his own company. He passed away in 2013.

Manlio Cavallini and his son Adrian J. Cavallini in their stained glass studio on Fredericksburg Road, San Antonio, Texas. (Monte Adams)

Manlio's son Adrian was brought up in the business and has been actively involved since he was in middle school. Today, Adrian's wife, Debbie, is an integral part of the business, responsible for the accounting, office tasks, and human resources. Their three sons, Joseph, Adrian, and Anthony, are third-generation Cavallinis that are showing interest in the business. Manlio says he knows they will continue to "run with the torch."

Adrian's Early Memories and Influences

Manlio's son Adrian, a San Antonio native, is deeply rooted in the community. His childhood memories, he says, include:

> *Myrtle and North Flores streets, right there by San Pedro Park. I remember going to the park and catching crawdads in the creek over there. But, probably one of my fondest memories etched in my brain is working with Tom Stell* [a famous muralist and painter who died in 1981]. *He was a real character and a great artist; he knew his mosaics and he knew how to do design. He would always make comments regarding the design. Or one of his craftsmen would be working and he would direct them how he wanted it done and the reason why. He would get upset sometimes.*

Standing in front of the stained glass window of the new Scobee Education Center in 2014 is the family of Adrian Cavallini: from the left is Joseph M., the middle son; Adrian J.; wife Debbie; Anthony, the oldest son; and Adrian V., the youngest son. (Photo Courtesy Adrian Cavallini)

Mario Scorza, a painter, was another designer that worked at The Cavallini Company. He was more of a mosaic artist and also a stained glass designer. Ann Norton came along and she was another good designer. The company also worked with other well-known designers such as Jan Howi, Seymour Fogel, Cecil Casebier, G. E. Mullan, Bro. Cletus Behlmann, and Paul Maxwell. Now, Roch Babin, whom we consider to be the best stained glass designer in the Southwest, has been with us since 1982. His talent and dedication are a big part of our success. So, with all these designers, I've been able to learn something new from each of them.

Growing the Business: The Early Years

Manlio joins in the conversation.

We started the stained glass about nine years after we began, but, at first, it was slow going. I would say until about the mid-'70s that's when we really began to take off. The first eight or ten years we were barely making a living.

Bell tower, University of the Incarnate Word. Stained glass restoration and protection glass by the Cavallinis.

(Don Kirk)

Manlio Cavallini next to a rack of stained glass samples. (Barbara Dean Hendricks)

Self-Taught

Manlio recalls:

> *I kind of taught myself the basics. I was smart enough to hire people who could do better than myself, but I learned all the fundamental processes so to speak. With the stained glass I started with very, very simple designs, but good designs, then we got into more and more elaborate designs and we became locally known. As we got into church work, the whole field was completely different.*

Like his father, Adrian is modest about following in his father's footstep. He says his early forays were not always top-quality work.

> *It's a learning process.*

Manlio adds,

> *I could tell right from the beginning that there were a lot of things that could be better in one way or another so it was kind of a reasonably long process.*

Making Glass

In June 1971, after considerable planning, Manlio Cavallini decided to initiate a new business by building a glass factory named *Multicolor Glass, Inc.*, in the southeast part of San Antonio. He made his own glass from

Designer Roch Babin figuring materials for a new stained glass project. (Monte Adams)

scratch basically using silica sand, soda ash, lime, and different oxides to obtain different colors such as blue, crystal, opal, green, amber, purple, and ruby. He started by producing a unique gift line and later manufactured a full line of glass globes for lamps including a full line of decorated items. By 1973, the factory had a three-ton furnace and one continuous furnace. The company did very well, establishing sales with lamp manufacturers and distributors on the West Coast, Florida, Dallas, Chicago, and other cities.

> *Eventually, because I was so involved in my stained glass business, I decided to go ahead and sell the glass factory in December 1980 to an attorney who didn't know a lot about glass; perhaps he bought it more for speculation and probably for a tax write-off. Eventually, he decided to shut it down.*
>
> *I was not very happy about that because we had close to forty-five different molds and they were sold out a little here, a little there, and then everything disappeared including one annealing lehr* [an oven used to anneal glass] *and one decorating lehr.*

Meanwhile, the Cavallini stained glass business slowly continued to grow. Manlio says,

> *At the beginning, it wasn't easy since there was already a very established company here in San Antonio by the name of 'Orco.'*

They were dominating the industry as far as church work. First of all, they had all the advantage of long experience, and they had the advantage of connections to buy materials, as well as well-established customers.

Manlio says there were other challenges:

The supplies of glass were restricted to very few companies. Let's say I wanted to buy glass from a company like Blenko or Wissmach, they would not sell to us because here in San Antonio they had sold to certain companies for years and years and they would not sell to anybody else. That kind of restricted trade is no longer true. Now, everything is wide open. You can go to class or you can teach someone how to make stained glass; you can buy books and all that, but you couldn't do that in the beginning.

An Example

Adrian and Manlio describe the steps to accomplishing a job with the St. Michael the Archangel Catholic Church in McKinney, Texas.

We incorporated twelve existing leaded glass windows into the main worship space. The transept stained glass encompassed fifty sections. Located in the clerestory windows above the altar are eight windows with sixteen sections each—that's a lot of windows for sure.

First of all, once we are in contact with the church and we know we're going to do the project, we sit down with the church representatives and get some feedback as to what they want. Of course, sometimes they have a definite idea, but sometimes they don't.

So, based upon information gathered during the conferences, we are then able to create a

An elegant sample of Cavallini etched glass on an entrance door added in a 2007 renovation of the historic 1906 Fairmount Hotel. (Don Kirk)

(Don Kirk)

Stained glass window detail,
St. Joseph Catholic Church, San Antonio, Texas.
Stained glass restoration and protective glass by the Cavallinis.

> *design and present it to the church for approval. Once the colored design is approved, then we proceed to fabricate the art glass.*

Colors and Glass Sources

You cannot talk about stained glass without talking about colors. Adrian notes,

> *Of the dalles themselves, there are maybe 500 colors. Then in art glass we probably have close to 1,500 colors.* ["Dalles" are one-inch-thick slabs of cathedral glass.]

Manlio adds,

> *Between the regular sheet glass and the dalles, we carry a tremendous number of colors.*

Their glass sources include the major companies in the United States: Kokomo Opalescent Glass in Kokomo, Indiana; Spectrum Glass Company in Woodinville, Washington; *Wissmach Glass Company* in Paden City, West Virginia; and the Blenko Glass Company in Milton, West Virginia.

The Palette

Adrian explains,

> *We do have a kind of color palette we like to work with, but it all depends on what we're trying to do. Every job is different. If we're*

Mission San Antonio de Valero (the Alamo) immortalized in stained glass at St. Dominic Catholic Church, San Antonio, Texas. (Barbara Dean Hendricks)

St. Dominic Catholic Church, San Antonio, Texas.
(Barbara Dean Hendricks)

Installing a pane of stained glass, St. Dominic Catholic Church. (Barbara Dean Hendricks)

trying to blend in with the interior of a certain environment or maybe we're going to try to achieve a certain mood and design, that sometimes determines what colors we use.

The Cavallini "Style"

Manlio believes that you wouldn't recognize a Cavallini window so much because of the glass, or colors, but for the quality of the design.

Nobody can match those masters, but growing up in Italy gives you a tremendous appreciation for artwork and also makes you aware that you have a challenge. Here, no two jobs are the same, so every time we take a job it's a big challenge because it's going to be something different with different people, different situations. To be sure, we never find ourselves getting bored.

"I picked the skills of the trade up by myself and by seeing what other people were doing." —Manlio Cavallini

It Begins with a Design

The process itself is very important; however, the beginning of a stained glass project is the design. You have to be able to make a design and—though it's not done all the time—you should also provide how you're going to structure the window and the glass, because the way you divide the glass, the way you use reinforcing bars, the size of the panes—all that has something to do with the strength and structure of the window.

Working with a Pattern

The original design drawings will be enlarged to the actual size of the window to create a pattern for us in cutting the different shapes of glass. Then comes what they call color selection, color coding on each piece of glass, and then the glass is cut to the specified sizes.

The Process: Painting and Firing

The paint is supplied in a powdered form and mixed with water and gum arabic. To paint the glass, among other things, a variety of brushes are used, but the most important brush of all is the 'badger'—a wide, soft brush with hairs three or four inches long used to smooth a wash of paint to a matte, muted, even finish. The badger is also used for stippling.

When required, the next stage is usually the application of silver or yellow stain. In this process, a silver compound—generally silver nitrate—is painted on the outer surface of the glass. When it is fired, the silver reacts with the substance of the glass to give a transparent yellow color varying from pale lemon to deep orange.

After the glass has been painted, it is fired in a kiln—a flat kiln that is specifically designed for our glass and fires at around 1,150 degrees. The paints are permanently fused to the glass and will last for hundreds of years if done properly.

The electronic kiln setter is programmed to fire the glass at the correct temperature in a certain amount of time, then automatically shut off in order to anneal the art glass.

Completion: Frames and Structure

We primarily install the glass into aluminum frames, but when wood frames are used it would be into an existing church. In the newer churches, they normally do not have wooden frames. It is not practical, not budget-wise, and not a wise choice so they use mostly structural aluminum.

Durability: St. Mary's Catholic Church

The Cavallinis are very mindful of the need for durability. Manlio says,

We want our glass to be the most durable possible because you are putting it in places that are made of bricks and masonry and so forth that are going to last for hundreds of years.

A shelf full of full-scale pattern drawings.

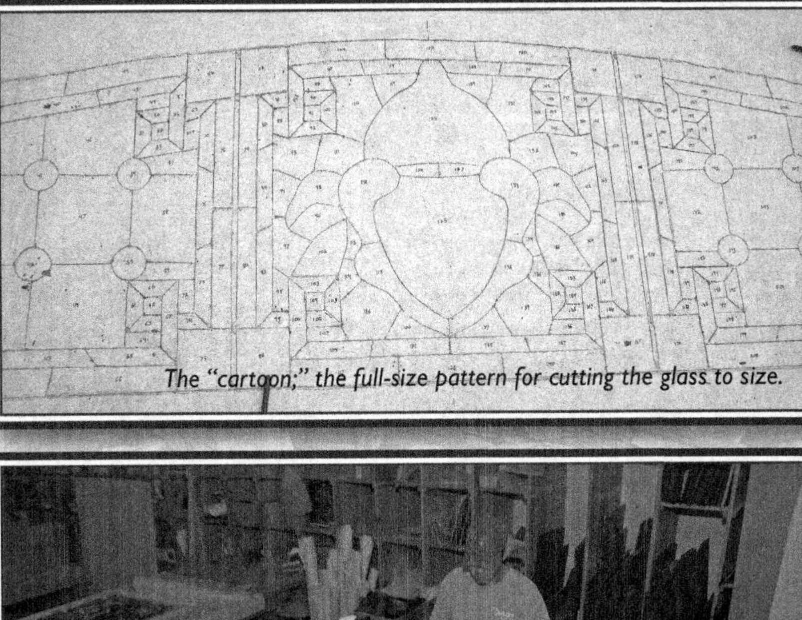

The "cartoon;" the full-size pattern for cutting the glass to size.

Glass pieces laid out and numbered.

(Don Kirk Photos)

Manlio & Adrian Cavallini & Sons, San Antonio studio.

Artist David Quintanilla applyin paint to art glas

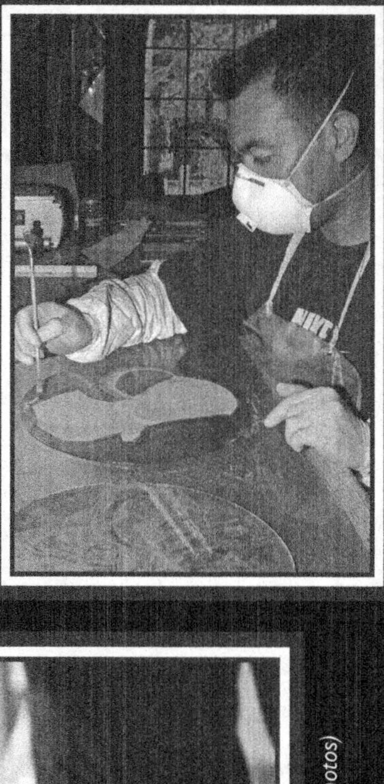

Designer Roch Babin directing his team.

Below: To form a faceted glass panel, one-inch-thick pieces of "dalles" are laid out in a frame and an epoxy resin poured between each piece of glass.

(Barbara Dean Hendricks Photos)

MANLIO & ADRIAN

oldering together the "cames" to orm a structural grid to support ndividual panes of colored glass.

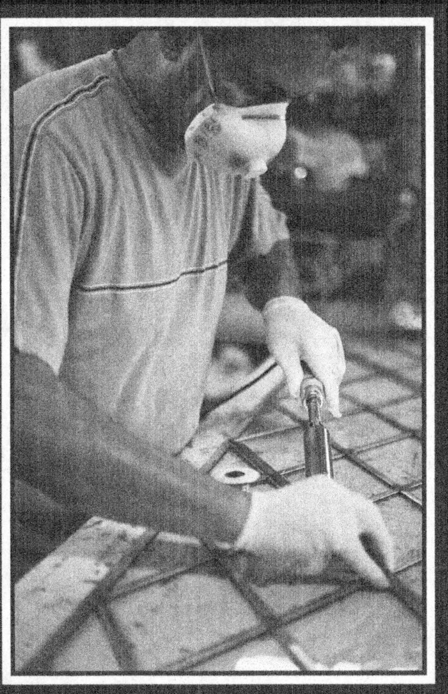

Carefully cutting glass with a glass cutter to match a full-scale pattern called a "cartoon."

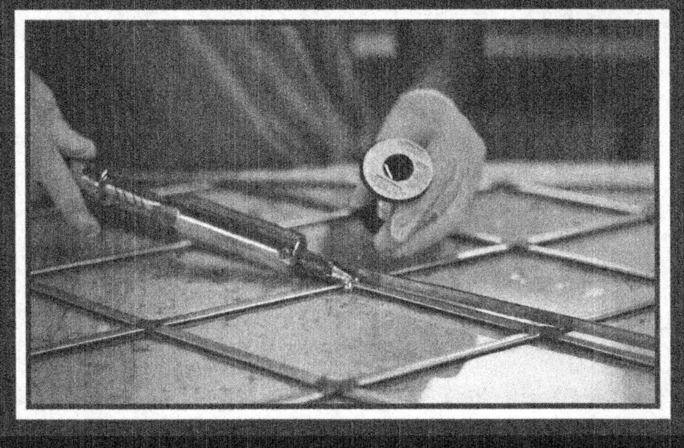

(Barbara Dean Hendricks Photos)

A soldering iron is used to heat the wire solder to its melting point to fix all the "cames" together.

CAVALLINI & SONS

Ryan Blackwell laying out the cut pieces of colored glass on the full-scale pattern in preparation for the "cames."

Missy Lopez working on a "cartoon."

Manlio Cavallini showing the cut glass laid out on the "cartoon" in preparation for assembly.

MANLIO & ADRIAN

Adrian V working on a stained glass project using the "cartoon" to lay out the pieces. (Courtesy Adrian Cavallini)

Carlos Jorda restoring an antique stained glass light fixture.

One-inch-thick faceted glass "dalles" in various colors are laid out on the pattern ready for a frame and epoxy resin.

(Don kirk Photos)

CAVALLINI & SONS

(Don Kirk)

BASILICA OF THE NATIONAL SHRINE OF THE LITTLE FLOWER
RESTORATION WORK BY THE CAVALLINIS.

Manlio tells about a recent purchase that truly illustrates the durability of stained glass.

> *We purchased all the stained glass windows from St. Mary's Catholic Church out of Port Arthur. All the windows are hand painted and beautiful. The church was heavily damaged from Hurricane Rita. It was beyond repair, so they tore the church down, but the windows survived.*

Adrian adds,

> *They were in some pretty substantial wooden frames that were still in good shape and they had protection glass on the outside. That helped to block the high winds. The church sustained heavy damage on the roof and subsequent water leaks in the walls, and it kind of deteriorated afterwards.*

Excellence in Design and Fabrication

High standards and quality control are the foundation of the Cavallinis' reputation. Adrian details some of the attention to detail required:

> *Like my dad was saying, before the inception of the design we account for the structural framework and all the mullions of the frames, and then we divide the panel up into structurally sound sections. We incorporate all the reinforcing bars into the design and we take into consideration all the structural elements.*

The work is built up from a core foundation, creating the quality and durability for which the Cavallinis are known. Manlio says,

> *All the materials must be of top-notch quality, and the craftsmen need to know what they are doing. They need to do good work and they need a fair amount of experience.*

Adrian details what that means.

> *They need to cut glass properly. They need to correctly pour castings for the faceted glass and apply the lead to the art glass. They need to do their soldering in a nice, uniform fashion. The waterproofing needs to be done right and cleaning needs to be done properly. Reinforcing bars must be properly soldered onto the stained glass panels in order to maintain their structural integrity. It's a never-ending attention to details. I do most of the inspection myself, and I also have other folks with more experience looking over the other craftsmen.*

Adrian at a layout desk in his studio. (Don Kirk)

Safety Concerns and Managing Hazardous Materials

Adrian has this to say:

> *We have safety programs and we've been very blessed not to have anybody with any problems. We abide by our safety rules and we enforce them, so if there's somebody that's a little sloppy we remind them what they're supposed to do and 90 percent of the time they respond. I maybe had to let one person go over all the years because he did not follow the safety rules close enough, but as a whole everybody follows them down to a 'T.' We keep everything real clean, plus we have all the proper equipment.*

The shop is organized and spotless. Adrian says,

> *We try our best, but it's always a battle.*

Manlio continues,

> *We also have more safety devices; we have filters; we ventilate the place; we use gloves; we check for lead content.*

Tools of the Trade

Times have changed and tools have changed since Manlio began.

> *Today, they teach you stained glass, like we also teach stained glass. We tell you what kind of tools you need, and there are all*

> *kinds of books you can buy. But when we started, back around the early 1960s, tools were not available, and—if they were available—we had no idea where to get them. For instance, a lead cutter that can be bought anywhere now, did not exist then. We had to make it ourselves. We didn't have to make every tool, but a lot of them. Those who worked in stained glass back then weren't going to tell you where to get it. It was kind of a trade secret.*

Stained, Faceted, and Beveled?

Adrian defines them:

> *Faceted glass is basically one-inch-thick 'dalle' that is about eight by twelve inches and is cut with a diamond blade to specific shapes that are then bound together with an epoxy resin. After the glass is cut, we take a carbide-tipped chipping hammer and just lightly chip around the edges of the glass to create the facets. A catalyst is mixed with the epoxy resin and then cast around each individual piece of glass at about 5/8″ thick. Approximately one-third of the window is epoxy. The weight is about eight to nine pounds per square foot.*

The 1929 Basilica of the National Shrine of the Little Flower, San Antonio, Texas. The Cavallinis did the stained glass restoration work. (Don Kirk)

BASILICA OF THE NATIONAL SHRINE OF THE LITTLE FLOWER
STAINED GLASS RESTORATION WORK BY THE CAVALLINI COMPANY, INCLUDING THE LARGE STAINED GLASS SKYLIGHT. (Don Kirk)

Etched glass creations by the Cavallinis, Pat Guidon Center, Oblalte School of Theology, San Antonio, Texas. (Don Kirk Photos)

Stained glass is basically 1/8″-thick glass bound together with lead, zinc, or brass. It's much more fragile and it's lighter of course. It weighs about four pounds or so per square foot.

The Sandblasting of Glass

Manlio explained another process that has been modernized.

We also do sandblasted glass. We use the word "sandblasting," but for health reasons we do not use sand anymore. Instead, we use aluminum oxide. Glass etching is an old method of imprinting images on glass. One way to do it is by starting from a light frosting to a multiple depth of carving, giving the finished product a richer, textured appearance, and a three-dimensional effect. We have done etching of designs, from a small piece to large areas as can be seen at Oblate Theological Seminary in San Antonio.

Specialty Glass

Manlio assures us that hand-blown glass is still being made, but some specialty glass from the past is harder to find. Lead glass, for instance, is still made, but in very limited amounts.

To match old opalescent glass is not as problematic since the 'Kokomo Glass Company' has been a longtime supplier.

Kokomo Opalescent Glass has been in existence from around the 1890s or so, says Manlio, and Adrian adds,

We probably have 300 to 400 colors, just from that factory alone, that we keep in stock. They are still in business and are a very reliable company, thank God.

Some textured glass such as Romanesque stained glass cannot be found. Texture can refer to glass with uneven thickness suggesting hand-blown glass. We can find some textured glass in old courthouses and other old buildings. I have searched for it high and low. There is some textured glass that just does not exist any longer.

The Artists

One important artist at the Cavallini Company today is Roch Babin, says Adrian.

He is pretty much responsible for all the artwork, and then we have several other people who do different things.

Designer Roch Babin working up a new stained glass window using his trusty colored pencils.

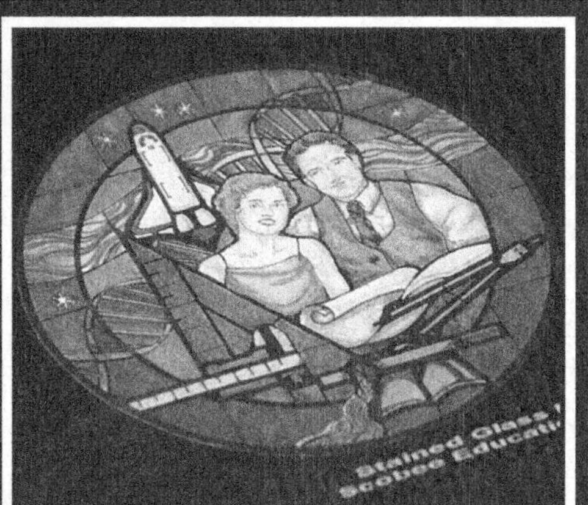

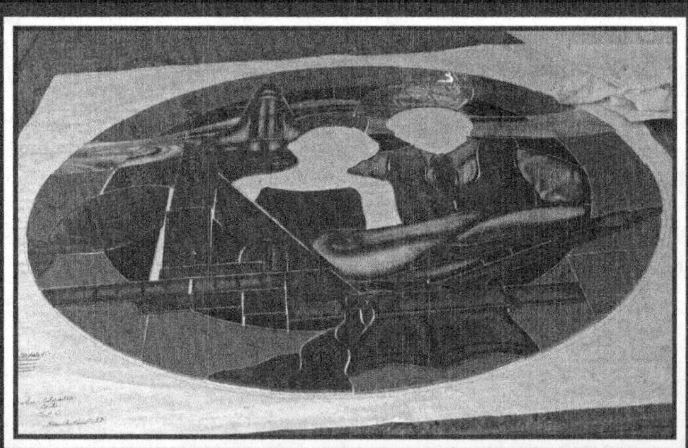

Above: Babin's colored-pencil study for the Scobee Education Center project. After approval, the concept goes quickly into production. At right: The cut out glass pieces.

Don Kirk Photos)

David Quintanilla painting portraits on art glass as part of a larger stained glass window for the Scobee Education Center at San Antonio College.

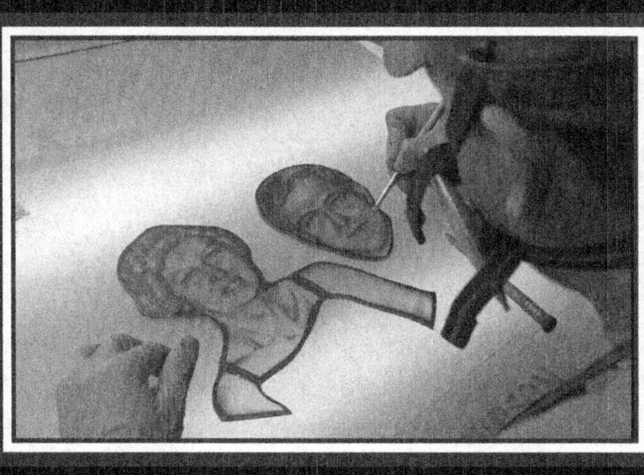

At left, detail of David's masterful work. These two glass pieces will fit in the pane shown at far left.

SCOBEE EDUCATION CENTER
(SAN ANTONIO COLLEGE)
The stained glass window of the new Scobee Education Center is dedicated to Dick Scobee (commander of the Space Shuttle Challenger which exploded after liftoff) and the astronauts who perished. The new education center was built around the old planetarium on the San Antonio College campus in San Antonio,
2014. (Courtesy Adrian Cavallini)

The two Cavallinis feel lucky to have Roch as an artist, Manlio says.

> *He has a degree in stained glass from LSU* [Louisiana State University], *a master's degree, and has been with the company since 1982.*
>
> *What happened was, about 1982, I had another artist who was a teacher—Ann Norton. She was pretty good, but she decided to go back to teach and I needed another artist. I couldn't find anybody around here, so I put an ad in the stained glass national magazine and I got three people who replied. One from New York; one from, believe it or not, Kerrville; and one from Roch Babin. He was in Kansas City at the time.*
>
> *He answered my ad and I called him over the phone and we had a conversation and I liked what he was telling me. I said, 'I'll tell you what, I'll fly up to Kansas City,' and I did. I saw his work firsthand and was very impressed with it. That's why I brought him down here. He's been with us ever since. I love his work.*

The Production Line

Adrian has this to say about production:

> *Back in the early '80s there might have been three of us. Now, we have about ten to fifteen people at any one time, so we became more compartmentalized. We had to kind of develop that over the years as we got more employees involved. We have an art department, the craft department, and so on. It's a process—our people are trained to work in certain areas.*

Manlio adds,

> *It's not something where you have one guy doing a design and that same person cuts out the glass and puts it together and that same person goes and installs it. You have to hire enough people and departmentalize it in such a way that production goes along reasonably smooth. The difference between us and Henry Ford and their production line is that they do the same product over and over whereas we have to custom design just about every product that we make.*

Workload and Customers

At the time of the interview, Adrian said they had probably a year's worth of work in line, but—due to their workflow process—they can accommodate new orders.

ST. MARY'S CATHOLIC CHURCH, SAN ANTONIO, TEXAS
RESTORATION WORK BY THE CAVALLINIS. (DON KIRK PHOTOS)

Compass-style mosaic on the San Antonio River Walk by the Cavallinis. (Don Kirk)

We work the smaller commissions in between the larger jobs. We do that constantly; otherwise we could never help anybody who has a smaller job. When I say 'smaller commissions' I mean like residential work. We do quite a bit of residential work and we just put it right into the production line with all the larger commissions.

San Antonio Historical Preservation

Manlio says,

I think from what I can see, the city is doing a very good job. Look at all these landmarks here in San Antonio: the San Fernando Cathedral, the Alamo, the missions, and, of course, the world-famous River Walk, where we recently installed a mosaic. There is a lot of history and art here because it's a multicultural city. The old art, the old buildings—they're very conscious whenever they make a change, especially in the downtown area. They make sure that they can blend in with the old architecture.

Adrian adds,

Sunset Station and St. Mark's Episcopal Church are good examples of Cavallini protection and restoration work. The Cavallini Company also did stained glass restoration and protection at St.

John's Lutheran Church downtown. They have some beautiful stained glass windows. San Fernando Cathedral, University of the Incarnate Word Chapel, and Our Lady of the Lake Chapel. That's all beautiful stained glass.

Cavallini Projects

Cavallini projects are not exclusively religious or restoration projects. The Cavallinis are quite active in commercial and contemporary work, including residential projects. Adrian says,

> *These clients usually want something kind of exclusive or something different, something unique.*

Manlio adds,

> *The Cavallini Company takes pride in the fact that we have done work for such prominent artists as Miguel Covarrubias, Carlos Mérida, Tom Stell, G.E. Mullan, George Pinca, Cecil Casebier, Paul Maxwell, Seymour Fogel, Bro. Cletus Behlmann, Ann Norton, and for famous people such as actor Thomas Gibson, Eva Longoria, Tony Parker, and Malik Rose.*

An example of the Cavallini Company's unique commercial work is Gallery Nord on Northwest Military Highway between Loop 410 and West Avenue. It's an art gallery with a very contemporary look that combines mosaic with stained glass. Manlio describes the work:

A rather unusual application with both stained glass and mosaics, Gallery Nord, San Antonio, Texas. (Don Kirk)

Sunset Station, formerly the Southern Pacific depot, San Antonio, Texas. (Don Kirk)

The Cavallinis fabricated a new sixteen-foot-diameter stained glass rose window for the north end of Sunset Station because the original had been removed by the previous owner. They also restored the matching south window. (Don Kirk Photos)

The all-business but cheerful Adrian Cavallini talking with a client. (Don Kirk)

> *If you go and look at the front of that gallery, you see several stained glass windows and then a continuation of that design going over the stucco wall with mosaics and then bringing it up to another stained glass window on the other side. It is the only one The Cavallini Company has ever done, and the company takes great pride in this unique project. It was the architect's idea—Allison Perry—it was his last project before he died.*

Dedication to Craft

Manlio Cavallini was eighty-one at the time of our interview; now he's eighty-six, but still works five or six hours whenever he comes to work, which is frequently. But, he said,

> *Let me make it very clear, Adrian is the one who carries the load. I still come in and help here and there. I was very much involved with the mosaics at St. Rose of Lima. I was there every day, on top of the scaffold.*

St. Rose of Lima

The St. Rose of Lima church is an excellent example of the diversity of the company. It is much more than solely stained glass. Manlio says,

The client wanted to put three mosaics on the bell tower, one on each of three sides. Roch designed them. Then we sent the design to Gino Barsanti, friend of mine in Italy, and he did the actual mosaic work and shipped the mosaic back to us for installation. A local tile company by the name of J&R performed the work under my supervision.

Custom Business and Marketing

Adrian notes that the business is just about 100 percent custom. They do very little advertising, relying mostly on references, recommendations, and their reputation for quality built over many years. Adrian notes they do have

a pretty nice-sized yellow page ad. We don't sit back behind the desk waiting for the phone to ring. We make phone calls; we go out and make site visits. We try to network as much as we possibly can and it works well.

The Competition

Adrian notes that their competition for contracts has greatly expanded over the years.

We compete on a statewide basis as well as nationally and even internationally, competing with studios from overseas.

Beautiful sample stained glass windows in Cavallinis office/studio. (Don Kirk)

ST. ROSE OF LIMA CATHOLIC CHURCH, SAN ANTONIO— ORIGINAL STAINED GLASS BY CAVALLINI & SONS.

(Don Kirk Photos)

BELL TOWER WITH LARGE MOSAICS ON THREE SIDES DESIGN, BY ROCH BABIN, MOSAIC WORK BY GINO BARSANTI— ST. ROSE OF LIMA CATHOLIC CHURCH.

(Don Kirk Photos)

Adrian V. Cavallini working on the restoration of a mosaic at Basilica of Our Lady of San Juan del Valle, San Juan, Texas. (Courtesy Adrian Cavallini)

Advice to Aspiring Artisans

> *When considering this field, you have to have artistic talent. You have to love art and then, if that's what you want to do, nobody will keep you from doing it. I mean anybody can really do it if you put your mind to it.*

Both men agree that the industry is in pretty good shape right now.

> *You can get a degree in stained glass, you can read books about it, and you probably can find out about it on the Internet.*

The Legacy: The Family Business

Manlio asserts,

> *The Cavallini legacy would be 'family.' They will be able to take over what I started a long time ago. You hear so many horrible stories where the founder of the company leaves the company and it crumbles. I don't think it's going to happen in my case. That's the legacy I want to leave to the family.*

Adrian had to make some real decisions. He majored in fine arts in college, but to be a designer is a full-time job, so you have to make that decision, 'Do I want to be a designer or do I want to have the knowledge of the art and also run the business?' In order to run a stained glass business, you have to have knowledge of the craft, of design, of the business world, everything. And you have to know how to sell your product. You have to wear different hats to address all those different facets of the business. If you don't, you're not going to make it.

Color rendering of the proposed design for Purple Heart mosaic. *(Don Kirk)*

They both agree there is never a dull day.

It's pretty exciting, always different, always a challenge.

Manlio Cavallini kneeling next to the finished Purple Heart mosaic at the Warrior and Family Support Center, Fort Sam Houston, Texas. *(Cavallini)*

DETAIL OF "THE LAST SUPPER" (Overall size: 27′ by 20′).

ST. JOHN NEUMANN CATHOLIC CHURCH, SAN ANTONIO, TEXAS.

TRADIT[illegible]NAL DESIGN BY THE CAVALLINI COMPANY. (Don Kirk Photos)

Original stained glass in the ceiling of the historic three-story lobby of the Menger Hotel, San Antonio, Texas. Protection and restoration by the Cavallinis. (Don Kirk)
Below: Archbishop's Residence, San Antonio, Texas. (Courtesy Adrian Cavallini)

Family Roles

Adrian recounts how the boys are choosing to learn about the business:

> *Joseph is learning the computer engineering and accounting side of the business. Adrian Vincent is mastering the fabrication and restoration of stained glass and mosaics, and Anthony is getting involved with public relations and marketing.*

The Cavallini Ethos

Adrian discusses some of the defining characteristics of the Cavallini beliefs and aspirations.

> *You get those values from your parents, obviously. They teach you to do things right—whether you work with them or not, whatever field you decide to go into. I was always brought up to put in a good honest day's work. Work hard—work smart, but work hard—and that's how you earn your money. I've kind of carried that throughout. It's really by action; you can say anything that you want to, but the proof is in the pudding, in the final product.*

Adrian applies this family philosophy to the family business every day:

> *You know, I'm back there in the workshop in the mornings. I spend most of my time overseeing what the workers are doing and then I go out on a job for the installation.*

New Technology and the Craft

Looking to the future, Adrian muses,

> *I think the further we get into technology, the more we want to go back to handcrafted mediums such as stained glass, painting, sculpture, or what have you.*

Manlio continues,

> *There are also companies that can develop the design out of the computer. We can just look at it and you can pick it up right quick, if you're aware of the topic.*
>
> *The images get recycled into other designs; you can see that they've just stretched it or made it more narrow. You know, the way you can do all that stuff on the computer. The computer is used for some aspects of the design or in the process of making the cartoons, but to totally rely on it 100 percent as your tool for design—well, that's not in our philosophy.*

Stained glass at St. Jerome Catholic Community, Houston, Texas.

Leaded glass at St. Jerome Catholic, Church, San Antonio, Texas.
(Photos Courtesy Adrian Cavallini)

Faceted glass of the creation, St. Mary's Catholic Church, Farmington, New Mexico.

The Cavalry of Christ, Faceted glass at St. Eugene de Mazenod, Brownsville, Texas.

First Baptist Church, Pasadena, Texas.

Stained glass detail, First Baptist Church, Pasadena, Texas. (Photos Courtesy Adrian Cavallini)

The Ancient Art of Staining Glass

Manlio takes on this question:

Stained glass is a very ancient art. Excavations in 1932 at Lorsch Abbey in Germany unearthed fragments of stained glass that are thought to date from the ninth or tenth century and believed to be the earliest existing pictorial stained glass. Germany has the distinction of having preserved the oldest complete windows in the world, made in the late eleventh century, in the cathedral in the ancient town of Augsburg, which was founded by the Romans in the first century AD–two thousand years ago. Stained glass as we know it today originated in Christian churches.

One of the special attributes of this art is its permanence and the meanings that resonate through time. Manlio says that durability is very important and believes that this kind of art may be appreciated in the future a lot more than it is right now. Here in the United States, he observes,

stained glass is not like in Europe, where it has been going on for centuries. This is a much younger art in this county, but people in America soon began to realize that stained glass is really a beautiful way to embellish both secular and religious buildings.

"It was for the people who couldn't read; that was the principal reason for stained glass." —Adrian Cavallini

Stained Glass as a Christian Art Form

Manlio reflects,

I think that people—whenever they go to church—they want to walk into an environment that has an atmosphere for prayer—there's a need for that. I think people, for the most part, are religious and they need to have something to relate to. That's where stained glass comes in. It started in the Christian churches. They used to explain the Bible through the figures and the symbols of the windows in the church.

Adrian adds,

It was for the people who couldn't read; that was the principal reason for stained glass.

AUGSBURG CATHEDRAL, BAVARIA GERMANY.
FIFTEENTH-CENTURY STAINED GLASS WINDOW BY PETER HEMMEL VON ANDLAU.
Manlio Cavallini's inspiration for traditional stained glass design.
(Photo Courtesy: Hans Bernhard–Schnobby–Wikimedia Commons)

Mosaics and etched glass in chapel of St. Monica Catholic Church, Dallas, Texas. (Photos Courtesy Adrian Cavallini)

Making Something Beautiful

Talking about what makes a beautiful window, Adrian declares:

> *I think it's a combination of things . . . If you look at the window and there's a very pleasant design and the colors are the right colors and the right proportions in the right setting—that is very important, the right setting—then you feel like you have accomplished what you set out to do right from the beginning.*
>
> *Of course, the other thing is, we do this work for all the different denominations and churches and our goal is to make them*

Etched glass divider panels and stained glass above them open into a large clerestory-lighted foyer. Below, an elaborate carved, etched glass mirror. Both can be found at the Dominion Country Club, and both are original designs by the Cavallini Company. (Don Kirk Photos)

CAVALLINI

A modern depiction of the crucifixion designed and fabricated by the Cavallinis in San Antonio, Texas. *(Don Kirk)*

Manlio Cavallini. (Don Kirk)

appreciate the art and appreciate their own beliefs. If we can do those two things, we have accomplished our goal.

Manlio smiles when asked about becoming a stained glass designer instead of a priest and says,

Well, yes, I mean, Adrian's here.

But, he affirms the positives about his original path:

I would never trade the seminary for anything else. I learned a lot about religion—the Scriptures—and about spirituality, and how to communicate with people and it really helps me a lot in the business I'm doing now.

Manlio has incorporated his religious studies into his artisan profession.

The intellectual side feels true to our minds and our souls.

The beautiful glasswork of the Cavallinis graces homes, schools, commercial buildings, and churches—inspiring, enthralling, and uplifting to all who experience it—so much more than just light through a window.

APPENDIX

APPENDIX

(Don Kirk)

OLD DOOR AT MISSION SAN JOSÉ, SAN ANTONIO, TEXAS

Glossary of Terminology

For Each Trade

A very condensed sampling of the language of our various artisans, but, nonetheless, interesting and educational reading.

1. The Woodcarver & Furniture Maker

(Johannes Scholze)

Back Bent Tool—as with "tracery" chisels (spoons), they are traditionally used for creating tracery features, but they produce a convex form on the inside of an internally curved surface.

Bleaching—a process that makes the wood's surface lighter than its original color and can help to remove unwanted marks.

Blocking In, Roughing Out—shaping the main forms and masses of a carving by chopping the bulk of the waste material away in the block of wood before the detailed carving begins.

Carver's Bench Screw—a screw with a tapered end for screwing into the bottom of the carver's work.

Carver's Mallet—used for striking gouges that have a rounded head.

Chip Carving—the technique of cutting out triangular-shaped chips to form decorative designs with a chip-carving knife. Used to cut out geometric patterns.

Chisel—a cutting tool with a flat blade.

Dog—a metal clip with serrated ends used with the aid of a screw for clamping work to a bench.

Facet—small, flat surface cuts that decorate the surface of a piece of wood. A combination of these cuts produces a "tooled" finish to carvings. A similar technique is used to decorate leatherwork.

Fishtail Tool—has the profile of a fish's tail. The lower part of the shank flares out in a triangular shape toward the cutting edge.

Flute—a groove having a curved section, especially when parallel to the main axis, as on columns and other cylindrical-shaped pieces.

Fluter Tool—similar to gouges except their cutting edge is not a uniform curve, but are "U"–shaped in section and used to carve flutes.

Gouge Tool—a chisel that has a curved cross section to its cutting edge.

Green Lumber—wood that is freshly cut and therefore not seasoned.

High Relief—a carving where the background is proportionately deep compared to the size of the carving. As a high relief gets deeper, the features approach three dimensions and may become isolated from the background.

Holdfast—a quick-adjusting clamping device that has a leg passing through a hole in the bench top. This allows work to be held in the middle of the bench, where clamps cannot be used.

Knife—a specialized tool used to pare, cut, and smooth wood. Various types include the mill, detailed, skew chisel, straight edge, roughing, bent, and the stab-and-cutting knife used for chip carving.

Low Relief—(also called "bas relief") a carving in which the background is proportionately shallow compared to the size of the carving, and the features defined by the actual physical depth of the background.

Maquette—a small-scale sculpture or full-size plasticine model made in preparation for the final piece.

Modeling—working the form of the surface.

Pierced—when a relief carving combines voids with cuts that go right through the design.

Quick Tool—a gouge with a deep curve to its cutting edge.

Re-Carving—to replicate a missing or broken decorative element by cutting material with sharp tools. If the carver uses period tools, the carving can be duplicated more accurately.

Relief Carving—the sculptural equivalent of a picture carving in which the depth is greatly reduced and compressed, but stands out from the background and has a limited view point.

Seasoning Timber—the process by which sap and moisture are removed from drying timber to reduce warping and increase stability by using one of two methods: 1) air drying: the wood is stacked and covered to protect it from the weather while only regulating the air flow in a natural way, but this is a slow process. 2) kiln drying: a kiln is used to completely control the wood's environment—the air circulation, temperature, and the humidity; this is quicker than air drying.

Sinking—reassessing the background so that the design stands out in relief.

Skew Tool—sometimes known as a "corner firmer," a chisel that has an angle to its flat-sectioned cutting edge.

Spade—these chisels are similar to "fishtail" chisels but have a longer elongated taper along their shank.

Spoon Gouge Tool—a hook-shaped tool with a uniform radius to the cutting edge, used to carve tracery.

Tooled Finish—a traditional technique that introduces facets on the surface of a carving, thus eliminating the need for abrasive materials. This gives life to the work.

Tracery—decorative openwork in the head of a Gothic window. Any delicate, interlacing work with branching lines in panels, screens, etc., as in carving or embroidery.

Undercut—to cut behind the edge of a design at an angle.

V-Tool—traditionally known as a "parting tool," as one of its functions was to separate or "part" the wood.

Whittling—the simplest form of carving, by using only a knife.

2. Paint & Finish Restoration Artists

(Cisi Jary & Daughters)

Age Crackle—a network of fine lines created by "embrittlement" of the entire finish layer (as opposed to "crazing," which develops in the surface of the finish layer). Continued expansion and contraction of the layer's substrate exerts a force that the aging finish layer can no longer accommodate as it "embrittles," resulting in the crackling of the finish.

Bole—a colored clay used as a preparatory layer for gilding; its color determines tonality of the gold.

Conservation—profession devoted to the preservation of cultural property for the future. Conservation activities include examination, documentation, stabilization treatment, and preventive care in accordance with an ethical code.

Consolidate—to glue down loose and flaking finish material.

Crazing—a microscopic network of fissures in an old finish layer (varnish, lacquer, or shellac) from age degradation. A hazy look resembling bloom, but "bloom" results from moisture while "crazing" is age related.

Degree—the comparative extent of damage described by adjectives such as "Negligible" through "Moderate" to "Extreme." Damage to a work of art is almost never static, and the work will continue to deteriorate over time.

Disfigurement—a change that defaces, modifies, or blemishes the appearance and intention of the original artisan. This may be intentional (as an element of a different style) or accidental (as with stains).

Finish—the final surface appearance of an object. It may or may not have a surface coating such as wax, shellac, varnish, or lacquer.

Fresco—the art or technique of painting on a moist plaster surface with colors ground up in water or simply a lime-water mixture.

Upper stories of the 1929 Tower Life Building, San Antonio, Texas. The eight-sided, neo-Gothic brick and terracotta tower was designed by architects Atlee & Robert M. Ayres. A copper tophouse and 100-foot flag pole were restored to the building in 2010. (Don Kirk)

Gesso—a hard compound of plaster of Paris or whiting in glue, used in sculpture or as a base for "gilding" or painting on wood.

Gilding—the application of gold leaf to an object, usually wood or metal.

Gold Leaf—gold that has been beaten into a very thin sheet, used in "gilding."

Grime—surface dirt or dirt that has been embedded into the finish.

Ingild—the process of matching and applying gold to areas where the gilding is missing.

Inlay—a design, pattern, or piece of material set into the surface of something.

Inpaint—the process of matching and applying finish to areas where the finish materials are missing.

Linseed Oil—a pale yellow oil extracted from the seed of the flax plant, used in paint and varnish.

Marbling—a faux technique to replicate the look of marble using paint applied directly to a wood surface.

Microscopic Finish Analysis—a method of identifying the components and different layers of finishes using the information obtained under a fluorescent-light microscope.

Mold—a furry growth of minute fungal "hyphae" occurring typically in moist, warm conditions, especially on food or other organic matter, such as the dirt, oil, and pollen layers found on furniture and other artifacts.

Patina, Natural—a surface finish or appearance induced by environmental factors over a prolonged period of time, often called "age patina."

Patina, Induced—a surface finish or appearance artificially induced by chemicals or other means to create a finish desired by the artisan.

Pigment—a dry, insoluble substance, usually pulverized, to be mixed with water, oil, or another base to produce paint and similar products.

Polychrome—made or decorated in many or various colors. An object or a work composed of or decorated in many colors.

Reversibility—the ability to be removed completely without affecting the surrounding original materials.

Staining—accidental discoloration or spotting created by contact with a foreign material.

Tone—The particular quality of brightness, deepness, or hue of a tint or shade of a color.

Watercolor—a pigment for which water and not oil is used as the vehicle. Watercolor paint is transparent, and the white of the background (paper) underneath provides light that shines through the painting.

3. The Stuccoist & Lead Beater

(Thomas Michael Battersby)

Armature—a framework used by a sculptor to provide structure and stability to a sculpture being modeled in a plastic material such as plaster.

Bel Composto—an artistic theory developed by Bernini during the Baroque period. The theory involved unifying the arts of painting, sculpting, and architecture. Interactions between the various mediums evoke emotional, sensual, and intellectual responses from viewers.

Baroque—a style of architecture and decorative art that flourished throughout Europe from the late sixteenth to the early eighteenth century emphasizing dramatic, bold, curving forms and elaborate ornamentation.

Brown Coat—the second coat applied over the "scratch coat" of conventional plaster.

Casting—a reproduction technique that duplicates the form of an original, whether modeled, carved, or constructed, but also makes possible certain effects that are impractical in the other techniques.

Coffer—a sunken panel in the shape of a square, rectangle, or octagon in a ceiling, soffit, or vault. The strength of the structure lies in the framework of these coffers.

Console—an ornamental scroll-shaped bracket that appears to support the floor or overhang above.

Corbel—a corbel is an architectural bracket or block projecting from a wall and supporting (or appearing to support) a ceiling, beam, or shelf. A corbel can be made of wood, plaster, marble, or other materials.

Cartouche—a structure or figure, often in the shape of an oval shield or oblong scroll, used as an architectural or graphic ornament or to bear a design or inscription.

Cornice—A horizontal molded projection that crowns a building or wall. Also, the molding in a room between the wall and the ceiling.

Dentils—small, rectangular blocks under a classical cornice resembling a row of teeth.

Embellishment—detailing or ornamentation used to adorn or beautify architecture.

Gargoyle—a grotesque representation of a human or animal figure with open mouth projecting from the gutter of a building, for throwing rainwater clear of a building's wall and foundation.

Grotesque—a style of decorative art characterized by fanciful or fantastic human and animal forms often interwoven with foliage or similar figures that may distort the natural into the absurd, the ugly, or into caricature.

Medallion—an oval or circular design used as decoration resembling a large medal. It often bears a figure in relief, a portrait, or an ornament.

Metal Lath—a mesh of metal used to provide a base for plaster.

Mold—a cavity that imparts its form to a fluid or malleable substance.

Modeling Clay—clay used to make a preliminary sculpture from which a finished work can be copied, such as using it as a mold for pouring plaster.

Ornamental Plaster—architectural details molded in plaster and attached to a building's structure. Moldings, cornices, and relief ornaments are cast separately and then mounted into place.

Plaster—a mixture of lime, sand, and water, sometimes stiffened with hair or other fibers, that is applied to the surface of a wall or ceiling as a soft paste that hardens when dry.

Profile Tool—a piece of hard material, like plywood, tin, cardboard, etc., cut to the contour or outline of a desired shape, like a cornice, so that it can be replicated in plaster.

Rococo—a style of architecture and decoration, originating in France about 1720, that evolved from Baroque types and was distinguished by its elegant refinement in using different materials for a delicate overall effect and by its ornament of shell work, foliage, and so forth.

Rosette—a disk of foliage or a floral design, usually in relief, used as a decorative motif.

Scratch Coat—the first layer or application of plaster, intentionally made rough so that the next layer will adhere to it.

Sculpture—the art of producing, in three dimensions, representations of natural or imagined forms. Sculpture embraces such varied techniques as modeling, carving, casting, and construction.

Stuccoist—a craftsman who creates decorative sculpted surfaces and sculptural detail for walls and ceilings in stucco.

Stucco—a material made of an aggregate, a binder, and water. Stucco is applied wet and hardens to a very dense, solid material. It is used as decorative coating for walls and ceilings and as a sculptural and artistic material in architecture.

Trompe L'oeil—an art technique that uses realistic imagery to create the optical illusion that the depicted objects exist in three dimensions.

4. The Faux Bois Concrete Sculptor

(Carlos Cortés)

Acid-Etched Color (Staining)—color applied to cementitious mixtures through the application of specially formulated acids containing various mineral salts that chemically react with the free lime in cement, producing a wide range of permanent colors that lightly penetrate into the surface.

Aggregates—those particulate materials added to cementitious mixtures primarily as inert and non-reactive "fillers," such as sand, gravel and rock.

Air Pockets—cavities in the concrete that trap air and cause the steel to rust prematurely.

Armature—a hidden framework serving as a supporting structure for the material used to make a sculpture. In *faux bois*, it's usually wire mesh and rebar. The Statue of Liberty has an armature of iron.

Bond—the adhesion between concrete or mortar and reinforcements or other surfaces. Adhesion between cement paste and aggregates.

Cement—a building material manufactured by grinding calcined limestone and clay to a fine powder, which can be mixed with water and poured to set as a solid mass or used as an ingredient in making mortar or concrete.

Color Staining (Color Washing)—to color with a coat of penetrating liquid dye or tint.

Concrete—a mixture of cement (usually Portland), various aggregate fillers (sand, gravel, rock), and water that chemically hardens into an insoluble monolithic composite. The most common formulation is often referred to as a "standard 1-2-3 mix" and consists of one part cement, two parts sand, and three parts gravel.

Cure—the method by which sufficient internal humidity and proper temperature are maintained to ensure that the hydration process of the cement within the freshly placed concrete or mortar is uninterrupted to allow full and proper hardening to occur.

Dry Brushing—highlighting with an unsanded grout.

Dionísio Rodriguez—the very finest work ever executed in *faux bois* style was done by this man. Having moved from México to San Antonio, Texas, in the early 1920s, he produced many remarkable and very large-scale installations all around the United States. Today, this virtually unknown artist is finally gaining the recognition he deserves, with many of his works

The church of the 1756 Mission San Juan, San Antonio, Texas. (Don Kirk)

now officially listed on the National Register of Historic Places. He currently ranks a close second to the esteemed Frank Lloyd Wright as having created the greatest number of sites attributable to one person.

Delamination—separation or breaking apart of a laminate into layers.

El Trabajo Rústico—the Spanish term for the *faux bois* style of concrete sculpture that was developed in México. Translated as "the rustic work," it is often characterized by a more realistic look in both composition and coloring, as well as a more finely detailed finish than comparable European works.

Faux—artificial or imitation, false.

Faux Bois—pronounced "foe bwah," a French term meaning "false wood," the school of arts & crafts dedicated to imitating either wood or wood grains in various mediums. While wood finishers routinely forge the grain, colors, and appearance of finer woods like mahogany, rosewood, and oak onto lesser woods such as pine, "ferrocement" *faux bois* utilizes concrete, mortar, and Portland cement paste to sculpt three-dimensional representations of trees, wood, and wooden objects onto steel frames or armatures.

Freeze/Thaw Resistance—the ability of a specific cementitious formulation to withstand the effects of repeated freezing and thawing, which can destroy most concrete because of the expansion and contraction forces of moisture within the cured mass. Air entrainment and/or "densification" are typically employed to counter these effects.

Ferrocement—a system of reinforced mortar or plaster applied over layers of metal such as chicken wire or expanded metal mesh or closely spaced small-diameter steel rods such as rebar.

Hickey—a sturdy, steel hand tool for bending rebar. It consists of a handle thirty to thirty-six inches in length welded to a triangular head bearing three stout posts for leveraging the rebar into various shapes. The single post on the reverse of the head can be fitted into a hole in a work table, allowing it to rotate around that point for additional leverage and control.

Galvanized Steel Mesh—a woven steel fabric with a zinc film to provide long term (generally ten to thirty years) corrosion protection.

Grout—a thinner form of mortar used to fill spaces like cracks and crevices in masonry or for finishing details. Unlike other structural pastes such as plaster or joint compound, grout, when mixed and applied correctly, creates a waterproof seal.

Mortar—any of various bonding materials used in masonry, surfacing, and plastering, especially a plastic mixture of cement or lime, sand, and water that hardens in place and is used to bind together bricks or stones.

Patina—giving the work an aged look.

Plasticity—the property of freshly mixed concrete, cement paste, or mortar that determines its ease of molding or resistance to deformation.

Rebar—short for "reinforcing bar," steel rods wired or welded together to provide a structure (armature) for applying the concrete.

Texture—applying a surface to the final coat of plaster to create the appearance of something else, such as tree limbs. Common tools used to achieve a texture include cake decorator bags, stiff brushes, table forks, nails, trowels, and knives.

5. The Metalworker

(Othon Garcés)

Annealing—the controlled heating and cooling of a metal to remove stresses and to make it softer and easier to work with.

Angle Iron—an iron or steel structural member that has been cast, rolled, or bent so that its cross section is L-shaped.

Arc Welder—a type of welder that uses a welding power supply to create an electric arc between an electrode and the base material to melt the metals at the welding point. They can use either direct (DC) or alternating (AC) current, and consumable or non-consumable electrodes.

Bar Stock—metal bars of various lengths, made in flat, hexagon, octagon, round, and square shapes, from which parts are machined.

Bench Grinder—a small grinding machine for shaping and sharpening the cutting edges of tools.

Brazing—joining metals by the fusion of nonferrous alloys having a melting temperature above 800 degrees Fahrenheit, but below that of the metals being joined.

Cold Working—a type of metalworking done by subjecting metal to enough mechanical stress to cause plastic deformation, a permanent change in the metal's crystalline structure. It gets its name because it is done at temperatures below the metal's recrystallization point and alters the metal's structure through mechanical stress rather than with heat. The major cold-working operations can be classified basically as squeezing, bending, shearing, and drawing.

Drill Press—a drilling machine with a counterbalanced spindle that makes it possible for the operator to control accurately the rate at which the drill is fed into the work.

Ductility—the property of a metal that permits it to be drawn, rolled, or hammered without fracturing or breaking.

Etching—a process used to dissolve unwanted materials by using acids.

Extrusion—the shaping of metal by forcing it through a die.

Fatigue—the effect on certain materials, especially metals, as they undergo repeated stresses.

Ferrous—a metal alloy in which iron is the primary ingredient.

Forge—to form or shape heated metal by hammering. Also, the name of the unit used for heating metal, as the blacksmith's forge.

Hardenability—a matter's capability of being hardened by a heat treatment.

Hardness—the property of matter that gives it resistance when force is applied.

Malleability—the physical property that allows the use of hammering or rolling to change a substance's shape or measurements.

Peen—to work on a metal using the force of a hammer.

Rivet—a permanent mechanical fastener. Before being installed, a rivet consists of a smooth cylindrical shaft with a head on one end. On installation, the rivet is placed in a punched or drilled hole, and the tail is "upset," or "bucked" (i.e., deformed), so that it expands to about 1.5 times the original shaft diameter, holding the rivet securely in place.

Steel—any of various modified forms of iron, artificially produced, having a carbon content less than that of pig iron and more than that of wrought iron, and having qualities of hardness, elasticity, and strength varying according to composition and heat treatment. Types of steel: alloy, carbon, tool, stainless, pickling, and toolpost. Alloy steel is steel that has been combined with other elements to improve its mechanical properties. Carbon steel has been combined with other elements, carbon the most often used element. Tool steel is any of a variety of carbon and alloy steels that have been combined to make tools. Stainless steel is an alloy steel with ten percent chromium content.

Welding—a process for joining similar ferrous-based metals such as steel and stainless steel. The metals are joined by melting and fusing the base metals being joined with a filler metal. Welding employs pinpointed, localized heat input.

6. Master Stone Masons

(Curtis Hunt & Family)

Arch—a curved stone structure resting on supports at both extremities, used to sustain weight, to bridge, or to roof an open space.

Architrave—the member of an entablature resting on the capitals of columns and supporting the frieze.

Ashlar—masonry having a face of square or rectangular stones, either smooth or textured.

Basalt—a dense-textured, igneous rock relatively high in iron and magnesia minerals and relatively low in silica, generally dark gray to black.

Bluestone—a dense, hard, fine-grained sandstone or siltstone of medium to dark or bluish-gray color that splits readily along original bedding planes to form thin slabs. Bluestone is not a technical geologic term. It is considered to be a variety of flagstone, the thin, relatively smooth-surfaced slabs being suitable for use as flagging.

Border Stone—usually a flat stone used as an edging material. A border stone is generally used to retain a terrace or platform.

Brownstone—a sandstone of characteristic brown or reddish-brown color that stems from a large iron oxide content.

Brushed Finish—obtained by brushing the stone with a coarse rotary-type wire brush.

Building Stone, Natural—rock material in its natural state of composition

Cornice detail of the 1881 Albert Maverick Building designed by architect Alfred Giles in the style of the Second Eclectic Period. The Maverick family along with the San Antonio Conservation Society, saved the building from demolition in the 1980s. (Don Kirk)

and aggregation as found in a quarry and usable in construction as dimension building stone.

Capital—the culminating stone at the top of a column or pilaster, often richly carved.

Carving—shaping by cutting a design into the stone, sculpting.

Cavity Vent—an opening in joints of masonry to allow the passage of air and moisture from the wall cavity to the exterior, also weep hole.

Celtic Drystack—method of constructing a wall without the use of cashel or mortar. Fitting is achieved though shaping.

Cement Putty—a thick, creamy mixture made with pure cement and water to strengthen the bond between the stone and the setting bed. Also called cement butter or cement cream.

Coping—a flat stone used as a cap on freestanding walls.

Cobblestone—a natural rounded stone large enough for use in paving; the term used to describe paving blocks, usually granite, generally cut to rectangular shapes.

Colonial Drystack—a stone-fitting style employed by the American colonists to construct boundaries and livestock barriers as they cleared their fields of stone on or in the topsoil. Very little shaping is employed.

Cornerstone—a stone forming a part of a corner or angle in a wall. Also a stone laid at the formal inauguration of the erection of a building, not necessarily at a corner, usually incorporating a date or inscription.

Cornice—a molded projecting stone at the top of an entablature.

Course—a horizontal range of stone units running the length of the wall.

Cut Stone—stone fabricated to specific dimensions.

Dressed or Hand-Dressed—the cutting of rough chunks of stone by hand to create a square or rectangular shape.

Drystack Wall—a stone wall that is constructed one stone upon the other without the use of any mortar. Frequently used for retaining walls.

Efflorescence—a crystalline deposit appearing on stone surfaces typically caused by soluble salts carried through or onto the stone by moisture.

Entablature—in classical architecture, the upper part of an order, comprising architrave, frieze, and cornice.

Exposed Aggregate—larger pieces of stone aggregate purposefully exposed for their color and texture in a cast slab.

Face—the exposed portion of stone. The word "face" can also be used when referring to the edge treatment on various materials.

Fascia—a horizontal belt or vertical face; often used in combination with moldings.

Fieldstone—loose blocks separated from ledges by natural processes and scattered upon the soil. Applied also to similar transported materials, such as glacial boulders and cobbles.

Flagstone—thin slabs of stone used for flagging or paving walks, driveways, patios, etc. It is generally fine-grained sandstone, bluestone, quartzite, or slate, but thin slabs of other stones may be used.

Grain—the easiest cleavage direction in a stone. "With the grain" same as "natural bed." Also, the particles (crystals, sand grains) of a rock.

Granite—fine-to coarse-grained, igneous rock formed by volcanic action, consisting of quartz, feldspar, and mica, with accessory minerals. Granite-type rocks include those of similar texture and origin.

Hearth Stone—originally the single large stone or stones used for the hearth, now most commonly used to describe the stone in front of the fire chamber and extending on both sides.

Honed Finish—a super-fine, smooth finish, though not as fine as a polished finish.

Igneous—one of the three classes of rock (igneous, sedimentary, and metamorphic), solidified from the molten state, as granite and lavas.

Inscription—lettering cut in stone.

Jack Arch—one having horizontal or nearly horizontal upper and lower surfaces. Also flat or straight arch.

Villa Finale, 401 King William Street, was the last home of preservationist Walter Nold Mathis, who was instrumental in restoration of the historic area. (Don Kirk)

Joint—the space between stone units, usually filled with mortar.

Jointing Scheme—architect's drawing detailing dimensions, location, and configuration of marble units and joints as related to the structure.

Keystone—the last wedge-shaped stone placed in the crown of an arch and binding the whole together.

Limestone—a sedimentary rock composed of calcium carbonate. Includes many varieties: recrystallized limestones and compact, dense, relatively pure microcrystalline varieties capable of taking a polish.

Lintel—the block of stone spanning the top of an opening such as a doorway or window, sometimes called the head.

Machine Finish—the generally recognized standard machine finish produced by planers.

Mortar—a plastic mixture of cement, lime, sand, and water used to bond masonry units.

Out of Wind—to have the arris (the edge) of a stone not laid in parallel or perpendicular lines so that it has an irregular or rustic appearance.

Overburden—a layer of soil and loose stone existing above the bedrock.

Parging—damp-proofing by placing a coat of 1/2-inch setting mortar to the back of stones or the face of the back-up material.

Pilaster—an engaged pier of shallow depth. In classical architecture, it follows the height and width of related columns, with similar base and cap.

Plinth—the lower, square part of the base of a column. The base block at the juncture or baseboard and trim around an opening.

Plucked Finish—rough-planing the surface of stone or breaking ("plucking") out small particles to produce a rough texture.

Polished Finish—the finest and smoothest finish available in stone characterized by a high luster, and strong reflection of incident light, generally only possible on hard, dense materials.

Quarry—the location of an operation where a natural deposit of stone is removed from the ground.

Quarrystone—stone extracted from a quarry, as opposed to fieldstone that has been harvested from a soil's surface.

Quartzite—a compact, granular rock composed of quartz crystals, usually so firmly cemented as to make the mass homogenous. The stone is generally quarried in stratified layers, the surfaces of which are unusually smooth. Its crushing and tensile strengths are extremely high, the color range wide.

Quoins—stones at the corner of a wall, emphasized by size, projection, rustification, or by using a different finish.

Relief—an ornament or figure slightly, halfway, or greatly projected to give a three-dimensional effect.

Reveal—the depth of stone between its outer face and a window or door set in an opening.

Ribbon—narrow bands of rock differing to various degrees in chemical composition and color from the main body of a wall.

Rip Rap—irregularly shaped stones used for facing bridge abutments and fills. Stones thrown together without order to form a foundation or sustaining walls.

Roman Arch—a semicircular arch.

Rubble—a product term applied to dimension stone used for building purposes, chiefly walls and foundations, and consisting of irregularly shaped pieces, partly trimmed or squared, generally with one split or finished face, and selected and specified with a size range.

Rustication—chamfers or square sinkings around the face edges of individual stones to create shadows and to give an appearance of greater weight to the lower part of the building. When only the horizontal joints are sunk, the device is known as banded rustication.

Sandstone—a sedimentary rock consisting usually of quartz cemented with silica, iron oxide or calcium carbonate. Sandstone is durable and has a very high crushing and tensile strength and a wide range of colors and textures.

Shaped Stone—cut stone that has been carved, ground, or otherwise processed.

Sill—a flat stone used under windows, doors, and other masonry openings.

Slate—a very fine-grained metamorphic rock derived from sedimentary rock shale.

Soapstone—a massive variety of talc with a soapy or greasy feel used for hearths, washtubs, tabletops, carved ornaments, chemical laboratory counters, etc., and known for its stain-proof qualities.

Stone—sometimes synonymous with "rock," but more properly applied to individual blocks, masses, or fragments taken from their original formation or considered for commercial use.

Terrazzo—a type of concrete in which chips or pieces of stone, usually marble, are mixed with cement and are ground to a flat surface, exposing the chips, which take a high polish.

Tile—a thin, modular stone unit.

Tooled Finish—a surface treatment that has four, six, or eight parallel, concave grooves to the inch.

Veneer—a non-load-bearing facing of stone attached to a backing for the purpose of ornamentation, protection, or insulation.

Weathering—the natural alteration by either chemical or mechanical means of stone due to the atmosphere, surface waters, soil, ground waters, or temperature changes.

7. The Blacksmith

(Kurt Pankratz)

Alloy—a mixture of two or more metals (bronze is made of copper and tin). Alloys are not compounds, but mixtures of metals dissolved into each other and crystallized upon solidification. Alloys often have properties much different than their constituents, such as greater strength, hardness, or a different melting point.

Anvil—a heavy object of durable material that resists the inertia of a hammer or mechanical ram while supporting the material to be struck. The modern blacksmith's anvil is a sophisticated tool developed over a millennia of tool making. Like many Iron-Age tools, it has roots in the Bronze Age. The modern blacksmith's anvil has a hard, heat-treated steel face and a durable, soft iron or steel body. The average blacksmith's anvil weighs 120 pounds. Most anvils have a flat horizontal face and a conical horn or "beak," but there are also hornless designs such as the "sawyers" anvil. Most anvils have a round punching hole, and all modern anvils have a square tooling hole called a "hardy" or "hardie hole."

The cornice of the eight-story 1909 Col. Gibbs Building, San Antonio, Texas. The hotel has been restored and is now 'Hotel Indigo' on East Houston Street. (Don Kirk)

Apprentice—a novice or new student of a subject. The definition of "apprentice" has changed with time and varies with the place or occupation. In the past, it meant one who was legally bound to a master in the arts as a student for a prescribed period of time. In exchange for his labor, an apprentice was given an education in the craft and the basic tools of the trade.

Bellows—a device that produces a strong current of air when its sides are pressed together. Used by the blacksmith to add oxygen to the coals of a forge.

Blacksmith—the word "smith" is said to come from the word "smite" or "to strike," as with a hammer. "Black" refers to the color of iron and steel after it has been heated and cooled. The surface has a covering of black oxide called "scale." Thus the blacksmith is one who strikes the black metal or the maker of things from the black metal.

Brass—an alloy made primarily of copper and zinc with trace elements typically including silicon and iron. Brass is golden yellow and weathers to green but is corrosion resistant. It can be very ductile but is often used to make castings.

Bronze—an alloy made primarily of copper and tin with trace elements, usually silicon and iron. Bronze is one of the oldest of the known alloys and was the first commonly used metal strong enough for tools and weapons. Bronze is a reddish color and weathers to a green color but is corrosion resistant.

Browning—a process of controlled rusting used to produce a rust-resistant oxide finish. Historically, this was used by early gunsmiths and other craftsmen to produce a dull surface on guns, tools, and other items, but the iron-oxide finish requires oiling to prevent further rusting.

Case Hardening—a process by which-carbon is absorbed into the surface of a piece of low-carbon steel or wrought iron, producing a hard "case."

Copper—a red, ductile metal occasionally found in its native state. Copper is an excellent conductor of heat and electricity, second only to silver in electrical conductivity.

Coppersmith—one who works copper, brass, or bronze with a hammer.

Drawing—lengthening a piece of metal by pulling, stretching, or forcing through a metal die.

Drift—a tool used in blacksmithing to make holes the correct size and shape. A hole is punched or cut in the material and a drift is forced through that hole to enlarge it to a particular size and/or to change its shape.

Escutcheon—a shield or shield-like surface on which a coat of arms is depicted. Also a protective or ornamental plate around a keyhole, door handle, drawer pull, light switch, etc.

Farrier—a specialist in equine hoof care, including the trimming and balancing of horses' hooves and the placing of shoes on their hooves. Blacksmithing skills are needed to make and adjust the shoes, and knowledge of the anatomy and physiology of the lower limbs and feet is required.

Flame Hardening—a surface-hardening technique using an intense heat source followed by quenching in water.

Flux—a chemical that dissolves metal oxides for the joining of the metals by soldering, brazing, or welding. The flux may also protect the metal from further oxidation while heating.

Forge—a device or place to hold an intensified fire for the purpose of metalworking. Also, the place, building, or shop where a forge is used. A typical forge has a forced-air source such as a bellows or blower to intensify the fire, an enclosure to hold the fire, and a chimney or vent to expel the smoke. Fuels include charcoal, mineral coal, heating oil, diesel fuel, propane, or natural gas.

Fuller—the process of shaping hot metal using a special tool or using the long, narrow peen of a hammer

Goliad County Courthouse clock tower, Goliad, Texas. This 1894 brick and stone Second Empire courthouse was designed by Henry E. M. Guidon. The tower and turrets were destroyed in a 1942 hurricane and were finally replaced in 2003. (Don Kirk)

to force metal to each side of the blow. A number of fullering blows spread the metal much like a rolling pin on dough, except that it is a much slower process in metal.

Goldsmith—one who works gold and gold alloys with a hammer. Probably the oldest of the smiths, as gold is commonly found in nuggets ready to be worked.

Hardy—a short, square-shanked chisel that fits in the square "hardy hole" of an anvil. Some have thin blades for hot cutting, thick blades for cold cutting, or curved blades for rounding the heel of a horseshoe. The hardy hole is named after the tool.

Hot Rasping—using a coarse file on hot iron or steel. The steel is usually anywhere from a dull red to a high black heat and the filing chips often come off ignited.

Iron—the word "iron" is used loosely to describe anything made of cast iron, steel, or wrought iron. In blacksmithing, forging steel is often called "pounding iron." Pure iron is a soft, ductile metal. Unprotected, it rusts very rapidly.

Journeyman—one who has finished his education as an apprentice and is ready to be employed elsewhere. Historically the journeyman traveled to other jobs to learn what could be learned and then moved on until ready to become a "master." Many journeymen never become masters and simply remain skilled in their field.

Peen—small end of a hammer opposite the face. May be round (a ball peen) or wedge shaped.

Penny Weld—copper brazing similar to zinc welding. The part is heated and fluxed, then the copper sprinkled on around the joint. The part is heated until the copper runs and then is removed from the fire to cool. The source of the copper often being pennies, the word was applied to the method.

Pig Iron—a common term for cast iron or cast-iron ingots. The term originated from the practice of casting ingots in triangular section troughs cut in the sand floor of foundries. A central trough fed smaller molds on its side like suckling pigs, thus "pig iron."

Pritchell—the slender, rectangular punch used by farriers to hot punch the nail hole in a horseshoe. The small round hole in London and American pattern anvils is used for this punching operation and is called the pritchell hole.

Scale—in ironworking, particularly forging, scale is the dark, blue-black gray oxide that forms on the surface of the iron, steel, or stainless when heated to forging temperatures. Scale is hard and abrasive, causing tools to wear down fast.

Shrink Rule—a measuring rule or scale that has increments increased in size by the amount that metal shrinks when making castings.

Smithy—the place the smith or blacksmith works; a blacksmith shop is a smithy.

Stake—a small anvil usually furnished with a tang to enter a hole in a bench top, as used by tinsmiths, blacksmiths, etc., for light work, punching upon, etc.

Steel—iron with a small percentage of carbon. The amount of carbon determines the steel's hardenability. The more carbon, the harder the steel can be made by heating.

Swage—a tool that shapes metal into a particular form: flat to round, or round to diamond shape, for example. The swage can be driven into the metal, the metal can be driven into a swage, or hot metal can be placed between top and bottom swages to be shaped overall.

Temper—the hardness resulting from the manufacturing process, heat treating, aging, or working as in work hardening. In foundry sand, it is the level of moisture and clay that results in the ability of the sand to bond and hold its shape. In ferrous alloys, it is the hardness after heat treating.

Temper Colors—the colors produced by the oxidation of clean steel used to indicate temperatures during the tempering process.

Tongs—this tool, used to grip and lift objects, must be able to withstand high temperatures. Types include straight, box jaw, knee, horseshoe, nippers, short jaw, V-bit, wolf jaw.

Trip Hammer—A heavy, power-operated hammer that is lifted by a cam or lever and then dropped repeatedly to flatten metal.

Upsetting—in forging, the process of driving metal back onto itself to increase the section area and reduce its length by hammering or pressing on the end of a bar.

Wrought Iron—pure iron crystals with thin layers of siliceous slag resulting in graininess similar to wood. Lack of carbon makes wrought iron unhardenable but very ductile. The graininess requires special handling in forging, cutting, and punching. Wrought iron is also the description of decorative ironwork that is made of any metal including wrought iron, steel, cast iron, and aluminum. Also used to describe low-carbon steel pipe.

One Ten Broadway, originally the 1907 Moore Building designed by Atlee B. Ayres, is a six-story yellow brick building with red terracotta egg-art molding, lion's heads, swags, and medallions on Houston Street in downtown San Antonio. (Don Kirk)

8. Terrazzo Artisans

(Richard Di Filippo & Family)

Aggregate—granular material, such as sand, gravel, crushed stone, and iron blast-furnace slag, used with a cementing medium to form a hydraulic-cement, concrete, or mortar.

Art Marble—artificial marble, precast terrazzo.

Bonding Agents—materials generally applied to "thin-set" terrazzo, (for example, latex, epoxy, or other types of adhesives). Used to increase adherence of the terrazzo mix to an existing base slab, also know as primers.

Broken Marble—fractured slabs of marble that have not been crushed by a machine into chips.

Chips—marble granules screened to various sizes.

Color Pigments—inorganic matter used in the terrazzo mix to vary the color. A powered substance that, when blended with a liquid vehicle, gives the matrix its coloring.

Control Joints—allowance made by the use of a strip, saw cut, or device to allow for movement without damaging the terrazzo.

Curing—the maintaining of proper moisture and temperature conditions necessary for the normal hydration of Portland cement to take place.

Grout—the binder or a color pigment applied to the floor to fill the voids and pits after rough grinding.

Marble—a recrystallized limestone composed of grains of calcite or dolomite, or both, having interlocking or mosaic texture.

Matrix—the Portland cement and water mix or non-cementitious binder used to hold the marble chips in place for the terrazzo topping.

Monolithic—cementitious terrazzo topping applied to a structural slab.

Panels—the areas formed by divider strips. Allows the pouring of individual "panels" without having to do the whole floor at once. Also used as a decorative element and to separate colors.

Precast—fabricated in molds, and finished in a shop or factory, by a compression and vibratory method.

Screed—to level the top of a mortar bed with a wood or metal straight edge.

Sealer—a required protective coating or treatment that prevents foreign liquid or matter from being absorbed by closing the pores in the surface.

Surfacing—the grinding, grouting and polishing operations used to finish terrazzo.

Swirl Marks—when using fine polishing stones in the final stage of finishing the terrazzo, it is normal to expect swirl marks. These marks will disappear rapidly with normal maintenance procedures and foot traffic.

Terrazzo—a mixture of marble chips and other fine aggregates used to give a decorative, marble-like look to the surface of finished concrete.

Tesserae—thin slices of marble, colorful stone, or glass-like highly colored vitreous enamel material cut into squares or other shapes of any size. Used in mosaic work.

Topping—the wearing surface of the terrazzo floor.

Underbed—cementitious materials to support divider strips and the terrazzo topping.

9. The Woodworker

(John Hall)

Architrave—ornamental moldings around a door or window frame covering the joint between the frame and plaster.

Balustrade—a kind of low wall that is placed at the sides of staircases, bridges, etc., that is made of a row of short posts topped by a long rail.

Brace—a hand tool used to drill holes, having a knobbed handle on the top to which pressure is applied, and a U-shaped grip in the middle used to rotate the drill bit.

Bench Dog—a peg that goes into various holes of a workbench to hold material in place.

Close Grain—woods with very fine fibers of cells that are not visibly porous.

Crook—longitudinal bending to one side, caused by uneven seasoning or grain.

Crotch—the section of a tree where a branch divides from the trunk, or the trunk divides in two, typically an area of convoluted grain.

Crosscut—a cut made perpendicular to the grain.

Cup—transverse bending, convex or concave, usually predictable by looking at the curvature of the grain on the end of the board.

Dado—a slot made across the grain.

Face—when a board has one side that is wider than the other, the wider side is referred to as the "face" (as opposed to the "edge"). May also refer to the face that is to be visible in the finished item.

Fence—a flat and straight length of material—usually wood, steel, or aluminum—that provides a surface to work against and prevents the work being cut from sliding.

Finial—a terminating ornament, as on the top of a post, piece of furniture, newel post, gable, spire, pinnacle, etc.

Flitch—a board in which the round of the trunk is still visible; a rough-cut board. Also, the complete bundle of thin sheets of veneer after cutting, laid together in sequence as they were sliced or sawn.

Flush—lying even with any adjoining surfaces.

Glazer—a person who selects, cuts, installs, replaces, and removes residential, commercial, and artistic glass.

Gouge—a chisel-like tool with a curved cutting edge.

Grain—the longitudinal fibers in wood.

Green Wood—unseasoned wood; freshly harvested timber, usually with a high moisture content.

Grit—the grade of particles in sandpaper or sharpening stones that determines the aggressiveness of the cut.

Hand Plane—a tool to smooth and true wood surfaces, consisting of a blade fastened in a frame at an angle with hand grips to slide it along the board.

Hardwood—lumber from the group of trees with broad leaves and does not refer to the actual hardness of the wood.

Heartwood—the wood from the pith extending to the sapwood, darker in color due to gum, resins, and other materials, which make it less susceptible to rot.

Hold Down—a type of iron clamp, that fits into a hole in a workbench and is tightened or loosened by hammer taps.

Jig—a device used to hold work or act as a guide in manufacturing or assembly.

Jig Saw—a power tool that cuts by moving a blade up and down as it is guided through the cut.

Joiner—a type of a carpenter who cuts and fits joints in wood.

Kerf—the width of a saw cut, determined by the thickness and set of the blade.

KickBack—when a work piece is thrown back by a saw blade. It can be prevented by using anti-kickback devices on power tools.

Kiln—a heated chamber for drying lumber where the airflow, heat, and relative humidity can be controlled.

Kiln Dried—lumber dried in a kiln.

Laminate—a product formed by bonding layers together as in built-up beams or plywood.

Light—a framed section of a window or door that is usually filled with a sheet of glass or other transparent material.

Linseed Oil—an amber-colored, fatty oil extracted from the cotyledon and inner coats of the flax seed. The raw oil extracted from the seeds by hydraulic pressure is pale in color and practically without taste or odor. When boiled or extracted by application of heat and pressure, it is darker and has a bitter taste and an unpleasant odor.

Milk Paint—paint made with milk solids, chemically akin to casein glue, often the original finish on antique furniture.

Mission Style—a design that emphasizes simple horizontal and vertical lines and flat panels that accentuate the grain of the wood.

Miter Box—an apparatus to guide a saw to make crosscuts at an angle.

Miter Gauge—a guide with an adjustable head that fits in a slot and slides across a power tool table to cut material at an angle.

The 1891 Reuter Building, designed by architect James Wahrenberger, has cornices of carved stone. To preserve the facade it was donated to the San Antonio Conservation Society. (Don Kirk)

Molding (*Moulding*)—a strip of material with a profile cut on the facing edges, used for trim work.

Mortise—a cavity or hole cut to allow a tenon to pass through it to form a joint.

Muntin—a strip of wood or metal ,separating and holding panes of glass in a window or door.

Nominal Size—the rough-sawn dimensions of a piece of lumber used to identify it, but do not represent the actual measurements; for example, a "2 x 4" is actually 1-1/2″ x 3-1/2.″

Pith—the soft core in the center of a log.

Plain Sawn—boards are sliced from the log with the cut tangent to the growth rings.

Plane—also referred to as a bench plane, a tool for smoothing and flattening boards.

Plumb—a term used to describe something that is perfectly perpendicular to the earth relative to gravity. A plumb bob on the end of a string will give you a line that is plumb—straight up and down.

Proud—to protrude a wood piece of a joint above the surface so it is sticking out a bit, usually so it can be sanded flush later.

Quarter Sawn—boards that have been cut so that the wide surfaces are approximately 90 degrees to the annual growth rings. This type of cut reduces cupping of the boards.

Rabbet—a groove in the edge or face of a board. Usually the groove is called a "rabbet" when on the edge and called a "dado" when on the face.

Radial Arm Saw—a circular saw that runs perpendicular to the work piece on an overhead track, but the track mechanism can swing left or right to make miter cuts.

Raised Grain—the roughened condition of sanded wood when the hard late wood rises above the soft early wood when moisture is applied.

Rasp—a long and flat steel tool with raised teeth for shaping wood,. Some are rounded on one side.

Rip Cut—a cut that runs through the length of a board parallel to grain.

Rough-Sawn Lumber—boards that are cut, edged, and trimmed, but not run through a planer.

Router—a high-speed motor with handles and an adjustable base with a chuck that accepts profile bits to cut dados, rabbets, or shapes.

Sandpaper—a form of paper with an abrasive material affixed to its surface.

Sapwood—the lighter-colored wood on the outside of a log. This wood is more susceptible to rot than heartwood.

Sash—a frame holding the glass in a window, typically one of two sliding frames. A "double-hung" is a window where both top and bottom sashes can be opened.

Sawhorse—a trestle usually used in pairs to hold wood up off the floor for easier cutting.

Seasoning—the time it takes for wood to dry. Wood air-dried takes one year for every inch of thickness, thus one year is a season.

Shellac—resin flakes dissolved in alcohol used as a finish for wood.

Shy—term for an adjoining object being below the object it is next to. This means a little below the surface. See *Proud* and *Flush*.

Spline—a thin strip of wood fitted between two grooves to make a joint.

Square—a device used to lay out or test right angles with two arms at 90 degrees to each other, the longer and wider arm is the "blade," the shorter narrower arm the "tongue."

Story Stick—(or "Story Pole") is a scrap strip of wood used to record dimensions for a project on site. Then the dimensions are used in the shop to build the project. This method reduces the chance of error due to the misreading of numbers or incorrect addition.

Stile—the vertical member of a door framework attached to the horizontal rails of a paneled door.

Table Saw—a circular saw mounted under a table with height and angle adjustments for the blade, allowing cuts in various depths and angles.

Tack Cloth—a cloth permeated with a sticky substance to wipe up the dust from sanding when finishing a project.

Template—a pattern to guide the marking or cutting of a shape.

Tenon—a projection made by cutting away the wood around it to insert into a mortise to form a joint.

Tongue and Groove—a joinery method in which a board has a protruding tongue on one edge and a groove on the other. The tongue of one board fits into the groove of the next.

Try Square—a tool for testing or laying out right angles, usually consisting of a metal blade fixed at a right angle to a wooden handle.

Turning—an ornamental or functional part formed by rotating it on a lathe and shaping it with a chisel-pointed tool.

Twist—a longitudinal twisting of wood due to uneven seasoning or grain.

Varnish—a liquid preparation that dries to a hard, lustrous coating.

Veneer—a thin layer of expensive wood bonded to a thicker piece of cheaper plywood to give the appearance of the expensive wood but at a reduced price.

Warp—a deviation from flatness as a result of stresses and shrinkage from the uneven drying of lumber. Types of warping include bow, crook, kink, cup, and twist.

Witness Marks—A remnant of original surface or scribed line, left during machining or hand working to prove that a minimum quantity of material has been removed or an outline accurately preserved.

10. Leaded, Faceted, & Etched Glass Artisans

(Manlio & Adrian Cavallini & Sons)

Acid Etching—a method of removing layers of flashed glass with hydrofluoric acid.

Annealing—cooling the formed glass product at a controlled rate of temperature change for the purpose of relieving thermal stresses.

Antique Glass—refers to a traditional method of manufacture, not its age. Antique glass is mouth-blown to eventually form a sheet when cut open and then allowed to flatten in a kiln.

Art Glass—specialty glass manufactured for use in stained glass projects and restoration.

Badger—a broad brush traditionally made of badger hair used to spread glass paint evenly.

Beveled Glass—a technique whereby thick pieces of clear glass are miter-cut and polished to create a prismatic effect. Bevels are available in a variety of sizes, shapes, and geometric configurations for use in leaded glass work.

Came—extruded, cast, or milled H-profile metal strip, usually made of lead (sometimes of zinc, copper, or brass), used to hold pieces of glass together to make a stained glass window. Individual pieces of came are soldered to each other where strips meet.

Cartoon—a full-size design drawing for making a stained-glass window or panel.

Cathedral Glass—machine-rolled clear or colored glass.

Cement—a particular kind of putty used in leaded glass. Cement is applied to fill the gaps between the glass and the channel of the lead came. This serves to both weatherproof the panel and add patina to the lead.

Copper Foil—copper-coated tape that is used to wrap around glass pieces providing a surface that can be soldered. Used as a method of building intricate windows and lampshades.

Copper Ties—short lengths of copper wire soldered to the interior surface of a leaded-glass panel. These wires are twisted around a steel "saddle" or round steel support bars to keep the window in place.

Crackle—a surface effect created when a hot cylinder of glass is dipped in liquid, causing radical, random fissures in the glass. The cylinder is then reheated and further blown to heal the surface fissures. The resulting surface resembles the texture of alligator skin.

Dalles—thick, usually one inch, slabs of cathedral glass.

Dalle de Verre—an art-glass medium in which dalles are broken into pieces with a carbon hammer and set in an epoxy base to adhere them in a decorative design. Dependent on large scale for best appearance, they are primarily used in architectural applications such as church walls.

Etched Glass—glass that has been engraved using an acid or other solvent to produce a frosted design on the glass.

Faceted Glass—glass panels that incorporate one-inch thick cast blocks called "dalles." Pieces are cut from these blocks and bound together with a matrix of poured epoxy to create a solid flat panel. The edges of the glass pieces are later chipped or faceted to create scalloped surfaces that capture light.

Flashed Glass—antique sheet glass that is made using two layers of different colors. Flashed glass is often used for etched or sandblasted glass art. When one layer is removed by etching, the second color is revealed.

Flux—a chemical agent (liquid or paste) used to facilitate the flow of solder and prevent formation of oxides during soldering.

Fused Glass—two or more separate pieces of glass that, when kiln-fired at high temperature, become permanently bonded.

Kiln—a high-temperature oven used by glass artisans to fire glass paint to glass surfaces, to cast or slump glass, or to fuse layers of glass together.

Leaded Glass—sheet glass pieces joined with metal strips, usually made of lead, called "came." Solder is applied to the joints of the came to bond the work together.

Lehr—a temperature-controlled kiln for annealing objects made of glass.

Mosaic—a picture or decorative design made by setting small, colored pieces of glass or ceramic material into a surface using cement or grout as a bonding agent.

Opal or Opalescent—describes any glass into which a material (usually fluorine or phosphorus) has been introduced at the raw-materials stage, causing a degree of crystallization and creating opacity in the glass. Reflected light is colorless—therefore white—and so white glass is commonly called "opal." Solid-color opalescent glass is glass that is both colored and crystallized, creating a single color sheet. Mixed opalescent glass is white glass (opal) mixed with one or more other colors to create a variegated, multicolored sheet. Light transmission varies with composition.

Painted Glass—glass on which special paints have been applied in a decorative pattern and then heated in a kiln to a temperature high enough to fuse the pigments permanently to the glass surface. It's the modern version of the original medieval "stained glass."

Patina—the chemical alteration of a metal surface resulting in a color change. Naturally occurring in lead when exposed to the elements, it can also be achieved more quickly by applying chemicals to the surface of the metal. Lead with a patina typically has the color and luster of graphite.

Pattern—a piece of a window design traced on and cut from paper. Pattern pieces are used to cut glass from a large sheet into smaller glass pieces of specific shape, size and color.

Protection Glass—a protective barrier installed over the exterior of a stained glass window. Commonly plate glass, laminated glass, tempered glass, or a polycarbonate, the protection glass is installed at least 3/4″ away from the leaded panel.

Rebar—round or flat steel bar soldered or wired to a leaded glass panel. Rebar (reinforcing bar) lends structural stability as it is then secured under the framing material holding it tightly in place.

Roundel—a glass panel, generally round, bearing a self-contained design.

Sandblasted Glass—a frosted dimensional effect created by masking off the surface of a piece of glass and spraying exposed areas of the design with an abrasive material.

Solder—an alloy of tin and lead, solder is heated with a soldering iron and used to join lead "came" or copper foil.

Stained Glass—commonly used to describe any colored, flat glass or any object made of such glass joined by metal strips. The term originally applied to colored or clear flat glass cut to fit an artist's design, on which details were painted in pigment with a brush. The glass pieces were then heated in a kiln or oven to bond the pigment to the glass surface. This firing makes the painted detail as durable and permanent as the glass itself. Most religious windows from medieval times until this century were executed in this manner. The term came to be used first for any architectural application, then for any design in colored, flat glass. Now universally accepted as a convenient general term to define the art, the craft, and the industry.

Stippling—a method of shading by dabbing paint.

Textured Glass—glass whose surface has been imprinted with a texture or pattern at high temperatures while still in the molten or malleable state.

Zinc—a hard metal that is formed into rigid channels and used to build or frame leaded glass panels. Zinc is frequently used to build beveled glass panels.

Glossary of Terminology Acknowledgments

Woodcarver & Furniture Maker: Some entries reprinted from the Woodworkers Institute website: www.woodworkersinstitute.com (© Copyright 2011 GMC Publications LTD or licensed for use by GMC Publications. All rights reserved.)

The Paint & Finish Restorer: Some entries are reprinted by permission of the Fine Arts Conservancy, West Palm Beach, Florida. www.art-conservation.org

Stuccoist & Lead Beater: Some entries in this glossary are from Wikipedia.com and are released under the Creative Commons Attribution Share-Alike License (CC-BY-SA)

Faux Bois Concrete Sculptor: Reprinted with permission from "Cement, Concrete & Hypertufa Glossary," © Copyright 2006 Holland & Tucker, All Rights Reserved. www.thegardenartforum.com. Cement entry from: American Heritage® Dictionary of the English Language, Fifth Edition. Copyright © 2011 by Houghton Mifflin Harcourt Publishing Company. Published by Houghton Mifflin Harcourt Publishing Company. All rights reserved.

Metal Fabricator: Some entries are reprinted by permission of Engineers Edge, Monroe, Georgia. www.engineersedge.com. Rivet & Arc Welding entries are from Wikipedia.com and are released under the Creative Commons Attribution Share-Alike License (CC-BY-SA).

Stone Mason: This section of the glossary was derived from Marble and Stone Slab Veneer by James E. Amrhein and Michael W. Merrigan, published by the Masonry Institute of America, 1989. www.braensupply.com. A few entries were sourced from: http://www.stone-network.com (Copyright 2007-2015 Stonebtb.com All rights reserved). Others from The Stone Network (stone-network.com), United Kingdom.

Blacksmith: reprinted by permission of Jock Dempsey at www.anvilfire.com.

Terrazzo Artisan: Reprinted by permission of Tercon Systems, Howell, Michigan, www.terconsystems.com

Woodworker: Some entries in this glossary are from Wikipedia's "Glossary of Woodworking" at Wikipedia.com and are released under the Creative Commons Attribution Share-Alike License (CC-BY-SA).

Leaded, Faceted, & Etched Glass Artisan: Some entries are reprinted with permission of Spectrum® Glass Company, Woodinville, Washington U.S.A. Some entries of the glossary are reprinted by permission of the Azure Stained Glass Studio, Cleveland, Ohio.

Thanks to all who contributed suggestions and additions to the glossary.

(Don Kirk 1980)

BEFORE RESTORATION OF A PRIVATE RESIDENCE,
KING WILLIAM HISTORIC DISTRICT, SAN ANTONIO, TEXAS.

(Don Kirk 2014)

AFTER RESTORATION OF A PRIVATE RESIDENCE, KING WILLIAM HISTORIC DISTRICT, SAN ANTONIO, TEXAS.

(Don Kirk)

THE RESTORED MISSION SAN JUAN, SAN ANTONIO, TEXAS.

Buildings & Artisans

A Selected List of Their Projects

1. Johannes Scholze: Woodcarver & Furniture Maker

Original Work

Aurora Apartments (*Mantelpiece*)
Fort Sam Houston (*Officers' Club*)
H. B. Zachry Home (*Beamed Ceiling*)
Menger Hotel (*Lobby*)
O'Connor Ranch (*St. Dennis Chapel*)
San Antonio City Hall (*City Seal*)
Texas State Capitol (*Portrait Frame*)
USAA (*Gen. McDermott's Desk*)

Restoration Work

The Alamo
Col. George W. Brackenridge Villa, University of the Incarnate Word (*Staircase*)
Edward Steves Homestead (*King William*)
McNay Art Museum
Mission Concepción
Mission San José
Spanish Governor's Palace (*Door*)
St. Joseph Catholic Church

2. Cisi Jary & Daughters: Paint & Finish Restoration Artists

Restoration Work

Briscoe Western Art Museum
Edward Steves Homestead
Empire Theatre
Fort Sam Historic Theatre
Majestic Theatre
McNay Art Museum
Menger Hotel (*Lobby*)
Mission Concepción
Mission San José
San Antonio Express-News Building
San Antonio Fire Station No. 1

St. Joseph Catholic Church
Texas A&M University (*Scoates Hall*)
The Alamo
Tobin Center for the Performing Arts (*Loggia*)
University of the Incarnate Word Mother House Chapel

3. Thomas Michael Battersby: Stuccoist & Lead Beater

Original & Restoration Work

Aztec Theatre (*Sweet Grapes, Mayan Mask*)
Drury Hotel (*Lobby*)
El Progresso Theatre
Empire Theatre
Exchange Building
Frost Bank Building (*City Council Chambers*)
Hotel Havana
Majestic Theatre
Menger Hotel (*Lobby*)
Robert E. Lee Hotel (*Lobby*)
San Antonio Express-News Building (*Ceiling*)
Smith-Young Tower Lobby, Texas A&M
St. Joseph Catholic Church
St. Patrick Church
Texas A&M University (*Ceiling*)
Trinity University Chapman Center (*Fireplace*)

4. Carlos Cortés: Faux Bois Concrete Sculptor

Original Work

Landa Library Grounds (*Palapa*)
River Walk Museum Extension (*Grotto*)
River Walk Museum Extension & Lavaca Street (*Benches & Bus Shelter*)
Robert L.B. Tobin Park, Salado Creek (*Tables, Benches, and Entrance Marker*)

Restoration Work

Brackenridge Park (*Trellis Bridge*)
Hildebrand & Broadway (*Old Trolley Stop*)
San Antonio Zoo Entrance (*Pergola*)
Sunken Gardens (*Entrance*)

5. Othon Garcés: Metalworker

Restoration Work

Edward Steves Homestead (*Fencing*)
Mission San José
Mission San Juan
San Fernando Cathedral
Southwest School of Art
St. Anthony Hotel (*Staircase*)
San Antonio Botanical Garden (*Sullivan Carriage House*)
The Alamo
The Arsenal
The Witte Museum

6. Curtis Hunt & Family: Master Stone Masons

Restoration Work

Bexar County Courthouse
Casa Navarro, Texas State Historic Site
Edward Steves Homestead
Empire Theatre
Fairmount Hotel
Fort Sam Houston
Lydia Ann Channel Lighthouse
Pearl Brewery
San Antonio Museum of Art
San Antonio Botanical Garden (*Sullivan Carriage House*)
San Fernando Cathedral
Southwest School of Art
The Alamo
The Arsenal
Wharton County Courthouse
The Witte Museum

7. Kurt Pankratz: Blacksmith

Restoration Work

Arneson River Theatre
Galveston Railroad Museum
Gonzales County Courthouse
Institute of Texan Cultures
Karnes County Courthouse
La Villita Assembly Hall
La Villita Historic Arts Village
Majestic Theatre
Mission Concepción
Mission San Joséf
Mission San Juan
Old San Antonio Public Library (torchères)
Olmos Park (*Gas Street Lights*)
Our Lady of the Lake (*Clock Tower*)
Pre–Civil War Mountain Howitzer
Southwest School of Art
Sullivan Carriage House (*San Antonio Botanical Garden*)
The Alamo
The Witte Museum
Villa Finale (*Walter Mathis Home*)

8. Richard Di Filippo & Family, Venice Art Terrazzo: Terrazzo Artisans

Original Work

Alameda Theatre
Alamo Museum Gift Shop
AT&T Center
Aztec Theatre (*Calendar Design on River Walk*)

(Don Kirk)

UNIVERSITY OF THE INCARNATE WORD, SAN ANTONIO.

Bexar County Courthouse
Cadena-Reeves Justice Center
Christus Santa Rosa Hospital
Fort Sam Houston (*BAMC Purple Heart Floor Mosaic*)
Freeman Coliseum
Henry B. Gonzalez Convention Center
Hipolito F. Garcia Federal Courthouse
Majestic Theatre
Palo Alto College (*Flooring & Countertops*)
Randolph Brooks Federal Credit Union (RBFCU) ASCII
Recruit/Family In-processing and Information Center, Lackland Air Force Base
St. Anthony Hotel
St. Dominic Catholic Church (*Fountain Mosaic*)
The Witte Museum
Tower Life Building
Trinity University (*Coates Hall Floor*)
University Hospital
UTSA Biosciences Building (*Floor Mosaic*)
UTSA Academic Building
UTSA Science and Engineering Building
UTSA Student Center

Restoration Work

Texas State Capitol

9. John Hall: Woodworker

Restoration Work

Atascosa County Courthouse
Aztec Theatre
Basilica of the National Shrine of the Little Flower
Bexar County Courthouse (*Windows*)
Casa Navarro State Historic Site
Christ Episcopal Church
Col. George W. Brackenridge Villa, University of the Incarnate Word
Fairmount Hotel
First Presbyterian Church
Fort Sam Houston
Guenther House (Pioneer Flour Mill)
Hotel Havana
IGN Train Station (now VIA Metro Transit's Westside Multimodal Transit Center)
Incarnate Word Motherhouse Chapel
Karnes County Courthouse
Lambermont Mansion
Laurel Heights United Methodist Church
Lone Star Brewery
McNay Art Museum
Mission San José
Pearl Brewery
Royalty Coins Building
San Antonio Botanical Garden (*Sullivan Carriage House*)
San Antonio City Council Chambers
San Antonio Museum of Art

San Antonio Zoological Society (*Barn Project*)
San Fernando Cathedral (*Altar*)
Spanish Governor's Palace
Southwest School of Art
The Alamo
The Witte Museum
Trinity University

10. Manlio & Adrian Cavallini & Sons: Leaded, Faceted, & Etched Glass Artisans

Original Work

Antioch Baptist Church
Archbishop's Good Shepherd Chapel (*Leaded Glass Figures*)
Archbishop Gustavo García-Siller Residence (*Mosaic*)
Army Residence Community Chapel (*Leaded Glass*)
Baptist St. Anthony's Hospital, Amarillo, Texas (*Faceted Glass*)
Basilica of Our Lady of San Juan del Valle National Shrine, San Juan, Texas (*Faceted*)
Brooks Air Force Base Chapel
City of San Antonio River Walk (*Mosaics*)
Covenant Presbyterian Church, Sulphur, Louisiana (*Leaded Glass in Chancel*)
Dallas Theological Seminary, Dallas, Texas (*Leaded Glass Figures*)
Fairmount Hotel (*Etched Glass*)
First Baptist Church, Pasadena, Texas (*Leaded Glass*)
Ford, Powell, & Carson Architects (*Leaded Glass*)
Fort Sam Houston (*Purple Heart Mosaic*)
Gallery Nord (*Stained Glass & Mosaic*)
Grace Avenue United Methodist Church, Frisco, Texas (*Faceted Glass*)
Haven for Hope (*Leaded Glass*)
Luby's Cafeterias #2 & #12 (*Etched Glass*)
Mary, Queen of Peace Catholic Church, Bossier City, Louisiana (*Faceted Glass Figures*)
Nix Hospital Chapel (*Faceted Glass*)
Oblate Grotto Chapel (*Mosaics*)
Old San Miguel Mission, Socorro, New Mexico (*Leaded Glass, Protective Glazing*)
Our Lady of Atonement (*Leaded Glass Figures & Symbols*)
Our Lady of the Miraculous Medal Retirement Center, Texarkana (*Faceted Glass Symbols*)
Our Lady Queen of Peace, Wichita Falls, Texas (*Leaded Glass Figurines*)
Salesian Sisters of St. John Bosco (*Leaded Glass*)
San Juan Diego Friary, Albuquerque, New Mexico (*Faceted Glass, Etched Glass*)
Schoenstatt Shrine, Austin, Texas (*Leaded Glass, Protective Glazing*)
Scobee Education Center, San Antonio College (*Leaded Glass*)
South Texas Children's Home Jess Lunsford Chapel, Mineral, TX (*Faceted Glass*)
St. Ambrose Catholic Church, Houston, Texas (*Leaded and Faceted Glass*)
St. Dominic Catholic Church (*Stained & Faceted Glass*)
St. Eugene De Mazenod, Brownsville, Texas (*Faceted Glass*)
St. Jerome Catholic Church (*Faceted and Leaded Glass, Mosaics*)
St. Jerome Catholic Community, Houston, Texas (*Leaded Glass*)
St. Joseph Catholic Church, Honey Creek, Texas (*Leaded Glass*)
St. Mary's Catholic Church, Farmington, New Mexico (*Faceted Glass*)
St. Mary Magdalen Catholic Church (*Interior Mosaic*)
St. Monica Catholic Church, Dallas, Texas (*Leaded & Etched Glass, Mosaics, Restoration*)
St. Pius X, Weslaco, Texas (*Leaded Glass*)

Southwest School of Art (*Leaded Glass*)
Sunset Station (*New Stained Glass & Restoration*)
Temple to the Brave, Beaumont, Texas (*Leaded Glass*)
TMI (*Leaded Glass*)
Travis Park United Methodist Church
University of the Incarnate Word Bell Tower (*Faceted Glass*)

Restoration Work

Anson Mills Building, El Paso, Texas (*Commercial Leaded Glass Restoration*)
Basilica of the National Shrine of the Little Flower (*Stained Glass Restoration*)
First Protestant Church, New Braunfels, Texas (*Leaded Glass Restoration*)
Incarnate Word Motherhouse Chapel (*Leaded Glass Restoration*)
Our Lady of the Lake University Chapel (*Stained Glass Restoration*)
Salado United Methodist Church, Salado, Texas (*Leaded Glass Restoration*)
St. Andrew the Apostle (*Leaded Glass*)
St. Anthony Hotel, San Antonio (*Lamp Restoration*)
St. Gerard Catholic Church (*Leaded Glass and Frames Restoration*)
St. Joseph Catholic Church (*Stained Glass Restoration/Protective Glass*)
St. Mark's Episcopal Church (*Leaded Glass Restoration*)
St. Mary's University (*Leaded Glass Restoration*)
Texas Lutheran University Chapel of the Abiding Presence (*Leaded Glass Restoration*)
Zion Lutheran Church, El Paso, Texas (*Leaded Glass Restoration*)

Arneson River Theatre, La Villita Historic Arts Village, San Antonio, Texas. This picturesque outdoor amphitheater on the San Antonio River Walk—used for dance and musical performances—was built by the National Youth Administration from 1939 to 1941. (Don Kirk)

Artisans' Work by Location

Arneson River Theatre

Kurt Pankratz, Blacksmith

Bexar County Courthouse

John Hall, Woodworker (*Windows*)
Curtis Hunt & Family, Master Stone Masons
Richard Di Filippo & Family, Venice Art Terrazzo

Brackenridge Park, Sunken Gardens, & San Antonio Zoo

Carlos Cortés, Faux Bois Concrete Sculptor (*Bridge, Entrance, Pergola*)

Casa Navarro

John Hall, Woodworker
Curtis Hunt & Family, Master Stone Masons

Edward Steves Homestead (*King William*)

Othon Garcés, Metalworker
Curtis Hunt & Family, Master Stone Masons
Cisi Jary & Daughters, Paint & Finish Restoration Artists
Johannes Scholze, Woodcarver & Furniture Maker

Empire Theatre

Thomas Michael Battersby, Stuccoist & Lead Beater
Curtis Hunt & Family, Master Stone Masons
Cisi Jary & Daughters, Paint & Finish Restoration Artists

Fairmount Hotel

Manlio & Adrian Cavallini & Sons, Leaded, Faceted, & Etched Glass Artisans
John Hall, Woodworker
Curtis Hunt & Family, Master Stone Masons

Fort Sam Houston

Richard Di Filippo & Family, Venice Art Terrazzo (*Purple Heart Mosaic Terrazzo Floor*)
John Hall, Woodworker
Curtis Hunt & Family, Master Stone Masons
Johannes Scholze, Woodcarver & Furniture Maker (*Officers Club*)

Landa Library Grounds

Carlos Cortés, Faux Bois Concrete Sculptor (*Palapa*)
John Hall, Woodworker

Majestic Theatre

Thomas Michael Battersby, Stuccoist & Lead Beater
Cisi Jary & Daughters, Paint & Finish Restoration Artists
Kurt Pankratz, Blacksmith
Richard Di Filippo & Family, Venice Art Terrazzo

McNay Art Museum

Cisi Jary & Daughters, Paint & Finish Restoration Artists
Johannes Scholze, Woodcarver & Furniture Maker
John Hall, Woodworker

Menger Hotel Lobby

Thomas Michael Battersby, Stuccoist & Lead Beater
Cisi Jary & Daughters, Paint & Finish Restoration Artists
Johannes Scholze, Woodcarver & Furniture Maker

Mission Concepción

Cisi Jary & Daughters, Paint & Finish Restoration Artists
Kurt Pankratz, Blacksmith
Johannes Scholze, Woodcarver & Furniture Maker

Mission San José

Othon Garcés, Metalworker
John Hall, Woodworker
Cisi Jary & Daughters, Paint & Finish Restoration Artists
Kurt Pankratz, Blacksmith
Johannes Scholze, Woodcarver & Furniture Maker

Mission San Juan

Othon Garcés, Metalworker
Kurt Pankratz, Blacksmith

Pearl Brewery

Curtis Hunt & Family, Master Stone Masons
John Hall, Woodworker

Royalty Coins Building

John Hall, Woodworker

Robert L.B. Tobin Park, Salado Creek

Carlos Cortés, Faux Bois Concrete Sculptor (*Tables, Benches, & Entrance Marker*)

San Antonio City Council Chambers & City Hall

Thomas Michael Battersby, Stuccoist & Plasterer (*Ceiling*)
Johannes Scholze, Woodcarver & Furniture Maker (*City Seal*)
John Hall, Woodworker

San Antonio Express-News Building

Thomas Michael Battersby, Stuccoist & Lead Beater (*Ceiling*)
Cisi Jary & Daughters, Paint & Finish Restoration Artists

San Antonio Museum of Art

Carlos Cortés, Faux Bois Concrete Sculptor (*River Walk Grotto & Benches*)
John Hall, Woodworker
Curtis Hunt & Family, Master Stone Masons

San Fernando Cathedral

Othon Garcés, Metalworker
John Hall, Woodworker (*Altar*)
Curtis Hunt & Family, Master Stone Masons
Kurt Pankratz, Blacksmith

Southwest School of Art

Othon Garcés, Metalworker
John Hall, Woodworker
Curtis Hunt & Family, Master Stone Masons
Kurt Pankratz, Blacksmith

Spanish Governor's Palace

Johannes Scholze, Woodcarver & Furniture Maker (*Door Restoration*)
John Hall, Woodworker

St. Anthony Hotel

Othon Garcés, Metalworker (*Staircase*)
Kurt Pankratz, Blacksmith
Richard Di Filippo & Family, Venice Art Terrazzo

St. Dominic Church

Manlio & Adrian Cavallini, Leaded, Faceted, & Etched Glass Artisans
Richard Di Filippo & Family, Terrazzo Artisans

St. Joseph Catholic Church

Manlio & Adrian Cavallini & Sons, Leaded, Faceted & Etched Glass Artisans
(*Stained Glass Restoration/Protective Glass*)
Cisi Jary & Daughters, Paint & Finish Restoration Artists
Johannes Scholze, Woodcarver & Furniture Maker

St. Térèsa (Little Flower) Basilica

Manlio & Adrian Cavallini, Leaded Glass Artisans (*Stained Glass Restoration*)
John Hall, Woodworker

Sullivan Carriage House (San Antonio Botanical Garden)

Othon Garcés, Metalworker
John Hall, Woodworker
Curtis Hunt & Family, Master Stone Masons
Kurt Pankratz, Blacksmith

The Alamo

Othon Garcés, Metalworker
John Hall, Woodworker (*Sashes*)
Curtis Hunt & Family, Master Stone Masons
Cisi Jary & Daughters, Paint & Finish Restoration Artists
Kurt Pankratz, Blacksmith
Johannes Scholze, Woodcarver & Furniture Maker
Richard Di Filippo & Family, Terrazzo Artisans

The Arsenal

Othon Garcés, Metalworker
Curtis Hunt & Family, Master Stone Masons

The Aztec Theatre

Thomas Michael Battersby, Stuccoist & Lead Beater
John Hall, Woodworker
Richard Di Filippo & Family, Venice Art Terrazzo

The Witte Museum

Othon Garcés, Metalworker
John Hall, Woodworker
Curtis Hunt & Family, Master Stone Masons
Kurt Pankratz, Blacksmith
Richard Di Filippo & Family, Venice Art Terrazzo

University of the Incarnate Word

Manlio & Adrian Cavallini, Faceted Glass Artisans (*Bell Tower*)
Cisi Jary & Daughters, Paint & Finish Restoration Artists (*Motherhouse Chapel*)
Johannes Scholze, Woodcarver & Furniture Maker (*Brackenridge Villa, Staircase*)
John Hall, Woodworker

Villa Finale (Walter Mathis Home)

Kurt Pankratz, Blacksmith
John Hall, Woodworker

Financial Support

Special thanks go out to these contributors for their undying support of this project and their contribution to the architecture of San Antonio.

Emily Denman Thuss

Building Arts Group Project Manager

Emily Denman Thuss has been active in numerous San Antonio civic and volunteer organizations for over twenty-five years, dedicated to empowering women of all ages and backgrounds. She chaired the committee that created a Master Plan for the Monte Vista Historic District and was District 9 Commissioner on the San Antonio Zoning Board. She was Program Coordinator for the Community Leadership Institute (Enterprise Community Partners) and worked to develop neighborhood leaders and homeowner advocates in some of San Antonio's economically depressed areas. As a member of the San Antonio Housing Trust, she administered over a million dollars in affordable-housing grants and loans. She was also active in the San Antonio Coalition of Neighborhood Associations, organizing events with the city, media, and the community. Emily also led in initiating, planning, and securing funding for Habitat for Humanity's ReStore, an innovative non-profit reseller of building materials donated by local builders and suppliers to encourage recycling and make home improvement more affordable. A successful economic generator for the San Antonio affiliate, ReStore has expanded to all areas of the United States. Over the years, Emily has worked to help her community and promote historical preservation. She formed the Building Arts Group in order to pay tribute to some of those craftsmen who helped preserve our architectural heritage in South Texas.

San Antonio Chapter
American Institute of Architects

Torrey Stanley Carleton, Hon. AIA Executive Director

The American Institute of Architects (AIA) has been the leading professional membership association for licensed architects, emerging professionals, and allied partners since 1857. Its national headquarters are located in Washington, D.C.

With nearly 300 state and local chapters, AIA serves as the voice of the architecture profession and the resource for members in service to society. Its membership numbers more than 85,000 nationwide and more than 6,200 in Texas. AIA San Antonio was founded in 1952 and today is the fourth largest chapter of the AIA in Texas, serving 550 members.

AIA is a member-driven organization focused on elevating public awareness, advocating for the profession, and creating and expanding the sharing of knowledge and expertise to ensure a prosperous future for its members.

AIA's contract documents are the accepted industry standard. These forms and contracts define the relationships and terms involved in design and construction projects. Prepared by the AIA with the consensus of owners, contractors, attorneys, architects, engineers, and others, the documents have been finely tuned during their 120-year history.

Since its founding, the AIA has become a robust, focused organization providing guidance, service, and standards to architects around the world. The AIA continues to strive for quality, consistency, and safety in the built environment and to serve as the voice of the architecture.

San Antonio Chapter of Associated General Contractors

Doug McMurry, Executive Vice President

The Associated General Contractors of America (AGC) was founded in 1918 after a request by President Woodrow Wilson. Wilson recognized the construction industry's national importance and desired a partner with which the government could discuss and plan for the advancement of the nation.

Today, it is a national not-for-profit trade association of more than 25,000 firms, including suppliers, general contractors, specialty contractors, and service providers.

Members of the AGC are dedicated to skill, integrity, and responsibility. Operating in partnership with its ninety-three chapters, the association provides a full range of services.

The San Antonio Chapter of the AGC is a ninety-two-year-old organization serving the building construction industry. Its 330 member firms build hotels, offices, schools, stadiums, hospitals, retail centers, military facilities, etc.

The San Antonio Chapter affiliates with the other ten Texas building chapters under the auspices of the Texas Building Branch (TBB). Primarily, the TBB serves the chapters by providing state government–relations services.

The South Texas Building, San Antonio, Texas

This twelve-story, 1919 Beaux-Arts-style commercial building was designed by Sanguinet & Staats and was originally known as the Central Trust Company Building. The name was changed to South Texas Building in 1932. It is constructed of steel and reinforced concrete and is faced with green granite, brick, and terracotta. The building was placed on the National Register of Historic Places in 1982. It is currently occupied by Home2Suites by Hilton. (Don Kirk)

San Antonio Conservation Society

Sue Ann Pemberton, FAIA, President

The purpose for which the San Antonio Conservation Society is formed is to preserve and to encourage the preservation of historic buildings, objects, places, and customs relating to the history of Texas, its natural beauty and all that is admirably distinctive to our state; and by such physical and cultural preservation to keep the history of Texas legible and intact to educate the public, especially the youth of today and tomorrow, with knowledge of our inherited regional values.

Founded in 1924, the San Antonio Conservation Society is one of the oldest and most active community preservation groups in the United States. Beginning with efforts to prevent historic structures from being razed and to preserve such unique sites as the city's Spanish Colonial missions, the society has been responsible for saving most of the historic attractions that make San Antonio one of the top tourist destinations in Texas.

Alamo Hardwoods

JR Huebinger and Biff Huebinger

Alamo Hardwoods can trace its roots back to 1921 as the 'J.C. Stout Lumber Company' through original receipts and phone records. By 1946, when Melvin Huebinger joined the firm, it had already become a branch of the Alexander Schroeder Lumber Company.'

Melvin and his partners bought the company and in 1958 'Alamo Hardwoods' was born. The focus remained on eighty woods of the highest quality from the U.S. and around the world and creating custom-designed moldings.

Today, Melvin's son, Biff, and his grandson, Biff Jr., continue the tradition of variety and high quality established almost one hundred years ago. Biff earned a bachelor's and master's degree in forestry and Biff Jr. a bachelor's of architecture.

They bring their unique perspective to an ever-changing industry. Any given day can bring in clients working on private jets for royalty or heads of state, someone creating a table for seven presidents to sit at for a G7 conference, a sword used by one of the defenders of the Alamo, or a local neighbor working on her 100-year-old screen door.

The Huebingers have worked hard over the generations to bring together skilled craftsmen, some second-generation employees, to work and shape the precious and common materials. They work closely with the local schools of architecture to employ interns to bring a higher level of knowledge into the design studio and then into each custom project. The company goal remains to provide the highest level of service to the client with the best quality wood available.

ALAMO HARDWOODS, SAN ANTONIO, TEXAS.
Complete Hardwood Lumber Supplier and Mill Since 1921.
Post Office Box 5398, #1 Fredericksburg Road, 78201.

JR Huebinger and his father, Biff, in their sales room.

Alamo Hardwoods stocks custom moldings, hardwoods, and softwoods, both domestic and imported. Specialities include boards and flooring, antique timbers, countertops, mantels, and veneers. Design services are available. (Don Kirk Photos)

Oral History Participants

Johannes Scholze
Woodcarver & Furniture Maker

Cisi Jary & Daughters
Paint & Finish Restoration Artists
(restorationassociates.com)

Thomas Michael Battersby
Stuccoist & Lead Beater
(battersbyornamental.com)

Carlos Cortés
Faux Bois Concrete Sculptor
(studiocortes.com)

Othon Garcés
Metalworker
(garcesdesigns.com)

Curtis Hunt and Family
Master Stone Masons

Kurt Pankratz
Blacksmith

Richard Di Filippo & Family
Terrazzo Artisans
(veniceartterrazzo.com)

John Hall
Woodworker

Manlio & Adrian Cavallini & Sons
Leaded, Faceted, & Etched Glass Artisans
(cavallinistudios.com)

Acknowledgments

The Building Arts Group *(Emily Denman Thuss, Barbara Dean Hendricks, Tommy Adams, and Monte Adams)*

Emily Denman Thuss *(Building Arts Group Project Manager)*

Barbara Dean Hendricks *(Author and Interviewer)*

Tommy Adams *(Financial Support, Photography)*

Monte Adams *(Photography)*

Ann Willoughby Kaplan *(Promotional Materials, Editorial Consultant)*

Sarah Sparrow *(Transcriptions)*

Jim Guyon *(Website Design)*

Michael J. Smith *(Photography, Promotional Materials)*

Karen Record *(Editor, Research, Fact Checking)*

Don Kirk *(Book Design, Layout, Research, Photography, Author)*

Additional Photography

A.C. Shelton
Adrian Cavallini
Curtis Hunt
John Hall
Karl Scholze
Kurt Pankratz
Melinda Garcés
Pamela Rosser Jary and Elisa Nieto Jary
Richard Di Filippo
The Trustees of Columbia University
Thomas Michael Battersby
UTSA Libraries Special Collections/W. Eugene George
Wikimedia Commons

The Emily Morgan Hotel, San Antonio, Texas

Originally built as the Medical Arts Building in 1926, this thirteen-story steel frame and brick building with Gothic detailing was designed by architect Ralph Cameron. The San Antonio Conservation Society holds the facade easement of this historic building.

(Don Kirk)

CPSIA information can be obtained
at www.ICGtesting.com
Printed in the USA
LVOW13*1909060518
576233LV00001B/1/P

9 781941 516065